TRANSPORTATION ISSUES, POLICIES AND R&D



# JUST-IN-TIME MANAGEMENT FOR AIRPORTS

# TRANSPORTATION ISSUES, POLICIES AND R&D

Additional books in this series can be found on Nova's website under the Series tab.

Additional E-books in this series can be found on Nova's website under the E-book tab.

TRANSPORTATION ISSUES, POLICIES AND R&D

# JUST-IN-TIME MANAGEMENT FOR AIRPORTS

LOW SUI PHENG
FAISAL MANZOOR ARAIN
AND
WONG YAN FANG

Nova Science Publishers, Inc.
*New York*

For permission to use material from this book please contact us:
Telephone 631-231-7269; Fax 631-231-8175
Web Site: http://www.novapublishers.com



**LIBRARY OF CONGRESS CATALOGING-IN-PUBLICATION DATA**

Just-in-time management for airports / authors, Low Sui Pheng, Faisal Manzoor Arain, Wong Yan Fang.
p. cm.
Includes bibliographical references and index.
ISBN 978-1-61728-783-1 (softcover)
1. Airports--Management. 2. Just-in-time systems. I. Arain, Faisal Manzoor. II. Fang, Wong Yan. III. Title.
TL725.3.M2L69 2010
387.7'360685--dc22

2010026615

*Published by Nova Science Publishers, Inc. † New York*

# CONTENTS

# PREFACE

The ability to deliver excellent quality standards in the most productive manner possible is the hallmark of successful organizations in this age of intense global competition. Japanese businesses have been able to compete successfully in the world market in recent decades because of their total dedication to quality and productivity issues. This has been made possible in part by the guiding principles of the Just-In-Time (JIT) concept which many Japanese businesses subscribed to. The JIT principles include waste elimination, pull production system, uninterrupted work flow, total quality control, top management commitment, employee involvement, long term working relationships with suppliers and continuous improvement. Beyond Japanese businesses, the JIT concept was also found to have benefited organizations in a wide range of industries including those relating to the built environment. While the JIT concept has its origin in the manufacturing sector, and in particular, the automobile industry, the JIT principles are equally applicable for managing resources in the built environment in various capacities. These include the spatial layout and design of buildings to enhance quality and productivity outcomes. This book explains how the JIT principles can be adopted for the air travel industry with specific emphasis on the management and operations of terminal buildings in airports. This book is primarily built on the case study of an excellent international airport that has won numerous accolades over a sustained period of time. It details a step-by-step explanation of how the JIT principles can be applied from the moment the passengers arrive at the airport to the time the passengers depart from the airport. The study covers major processes and procedures typical of the spatial management and operations of major airport terminal buildings. The evaluation for JIT application includes the points of arrival and departure, the check-in hall, immigration area, transit mall, gate lounges, food and beverage outlets, retail shops as well as other management initiatives that strive for continuous improvement. The analysis shows significant potential in applying JIT principles for managing airport operations within the confines of the physical airport terminal buildings. Given that major international airports around the world are striving to compete for a market share of the global air travel industry, it is important for them to deliver excellent customer value that is anchored strongly on continuous quality and productivity enhancement. This book is a valuable resource for all airport managers, operators and design professionals to realize how this enhancement can be accomplished through the JIT concept.

Much of the analysis presented in the study would not have been possible without the valuable feedback, comments, suggestions, recommendations and assistance provided by the professional airport managers and airport staff in the fieldwork conducted in Singapore. It is their professionalism and dedication as excellent service providers that made this study possible. It is hoped that this book would, in turn, be able to offer all professional airport managers globally an alternative JIT-based best practice perspective to review, evaluate and improve on the management practices and operational work flows in airports worldwide.

Low Sui Pheng
Faisal Manzoor Arain
Wong Yan Fang

*Chapter 1*

# INTRODUCTION

## BACKGROUND

The Just-In-Time (JIT) concept was developed in Japan by the Toyota Motor Corporation in the early 1950s. The concept had provided a useful strategy which has enabled Japanese car manufacturers to compete successfully against American and European car makers globally since the end of the last World War. The JIT concept is a management philosophy which helps to integrate and streamline the manufacturing system into its simplest possible processes. It conceptualizes how the major resources for manufacturing (namely capital, equipment, materials and labor) should be made available only just-in-time for production and in the amounts required for completing a specific job. The JIT concept is a continuous process of striving to minimize elements in manufacturing systems which restrain productivity. When successfully implemented, the JIT concept will reap the following benefits: reduced inventory levels, improved quality levels, reduced scrap and rework rates, reduced manufacturing lead times, improved customer service levels, improved employee morale, and overall improvement of productivity and profitability. The seven fundamental JIT principles are shown in Figure 1.1. These are described briefly below drawing from examples in the prefabrication sector of the construction industry.

## WASTE ELIMINATION

The first JIT principle is "waste elimination". Waste in the JIT environment means anything that does not add value to the production process.

An appropriate and achievable goal in the prefabrication setting is waste minimization rather than total waste elimination. By adopting appropriate JIT based strategies, waste resulting from over-production, undesirable waiting times, excessive transportation and handling, processing operations, inventory build-ups and product defects can be minimized (Low and Chan, 1996, 1997).

Attacking fundamental problems is a key ingredient to accomplish waste elimination. Rather than masking problems, the JIT concept attacks the fundamental root causes that gave rise to the wastes, problems or defects. For example, identifiable problems in the prefabrication sector can include the "open factory" layout of most prefabrication operations which are exposed to adverse weather conditions, general construction problems, and labor intensive operations. By attacking these fundamental root problems, waste elimination can be accomplished. Quality and productivity enhancement as well as cost reduction can be achieved in the prefabrication sector that produces precast concrete components for the construction industry (Low and Chan, 1996, 1997).

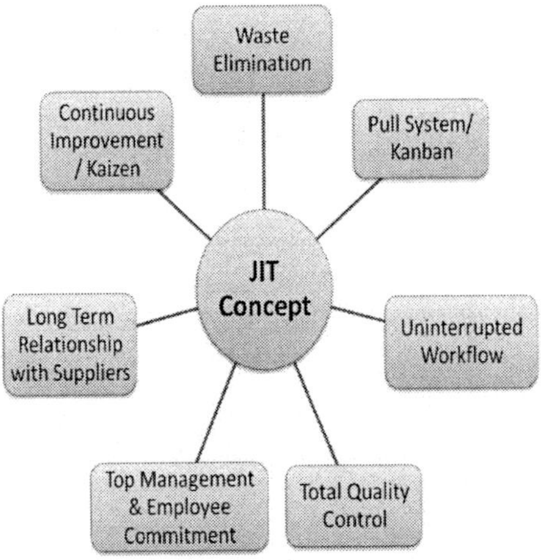


Figure 1.1. Seven fundamental principles of the JIT concept.

## KANBAN OR PULL SYSTEM

The second JIT principle is the use of the "kanban (card in Japanese) or pull system". This refers to an information processing system controlling the production cycles in which necessary products are produced just-in-time in the exact quantities that are required for the various production processes. Buffer stocks are thus minimized or eliminated between the various production stages. As an example, while the prefabrication sector is already utilizing the kanban system in the production of non-standardized one-off precast concrete products, the system can, in fact, be extended further to the production of standardized, commonly used precast concrete products as well (Low and Chan, 1996, 1997). Hence, the kanban system can be integrated with existing manufacturing resource planning systems for effective forecasting, planning and control of the production of all precast concrete products. However, to achieve this, certain structural modifications would first need to be made to improve and maximize the operating efficiencies of all sub-processes. In this context, the kanban system would seem to be most suitable for application in materials management in the prefabrication sector of the construction industry (Low and Chan, 1996, 1997).

## Uninterrupted Workflow

The third JIT principle is "uninterrupted workflow" which requires a smooth flow of work processes. The production process should be rationalized and simplified as much as possible. In the prefabrication sector of the construction industry, uninterrupted work flows can be achieved by utilizing "group technology" techniques to rearrange interrelated production processes in a "cellular" arrangement, designing objectively for simplification, optimizing the usage of mechanization and automation, and encouraging reduced process set-up times through the practice of small lots production (Low and Chan, 1996, 1997).

## Total Quality Control (TQC)

The fourth JIT principle draws on the "total quality control (TQC)" concept. No JIT system can function without high quality parts and products. As buffer stocks to allow for defective parts are not carried in a JIT production system, quality problems must be resolved at individual work stations, which shift the responsibility for quality control to the individual makers of parts. In this context, the prefabrication sector in the construction industry has made significant progress towards quality certification to meet ISO 9000 requirements. The TQC concept which embraces strategies such as quality checking at source, quality control circle improvement teams, and effective total quality management can help to further fine-tune the quality certification process in the prefabrication sector (Low and Chan, 1996, 1997).

## Top Management Commitment and Employee Involvement

"Top management commitment and employee involvement" is the fifth JIT principle. The JIT approach requires fundamental and radical changes both to management functions and the way in which labor is organized, trained and rewarded. Top management must be willing to accept changes such as stockless production, implementation of a total quality management system, increased delegation of responsibilities to employees, single source suppliers and adopting the kanban/pull system. On the other hand, employees should be ready to accept changes such as increased levels of responsibilities, positive working attitudes, continuous education, work participation as well as perform as multi-functionally trained workers. Existing labor problems relating to the single-skilled workers and the generally high level of illiteracy among workers in the prefabrication sector also lend themselves to the application of the JIT principle to enhance labor productivity (Low and Chan, 1996, 1997).

## Long Term Working Relationships with Suppliers and Clients

Fostering "long term working relationships with both suppliers and clients" is the sixth JIT principle. In the JIT system, suppliers are singly sourced and are expected to practice JIT principles to ensure long term high quality defect-free supplies. They are also expected to deliver such supplies in smaller lots at frequencies specified by actual production rates downstream. In the context of the prefabrication sector, clients are also expected to furnish accurate schedules and expedite co-ordination to ensure the smooth production and delivery of precast concrete components for just-in-time installation on site. Through the promotion of an industry-wide JIT movement and the co-operation of both the suppliers and the clients, the prefabrication sector can adopt JIT-based strategic actions to improve productivity in the sector. For example, long term purchaser agreements and single-source supplier networks can be implemented readily, given the comparatively small number of production sub-components required in the prefabrication business (Low and Chan, 1996, 1997).

## Continuous Improvement or Kaizen

The seventh JIT principle relates to "continuous improvement or kaizen". The apparent lack of a productivity performance assessment system, the low degree of standardized production, the insignificant level of automation used, and numerous unresolved technical problems are some of the issues that can be addressed in the prefabrication sector by directing efforts towards continuous improvement (Lim and Low, 1992).

These seven JIT principles will be elaborated in greater details later.

## JIT and Airport Management

Various studies have suggested that the implementation of the JIT concept can help to raise operational and business efficiency. In this context, the JIT concept has been adopted in diverse industries that ranged from manufacturing to construction. By extrapolation, it is possible and desirable for the JIT concept to be extended to other industries, including the aviation industry, to help raise quality and productivity standards as well as to meet customer satisfaction.

From the point of view of the global aviation industry, the growth of the Changi International Airport in Singapore has both been healthy and phenomenal. For example, in the period between 2002 and 2007, there was an increase of about 26% in the number of scheduled weekly flights operating out of the Changi International Airport. The total number of passengers who used the Changi International Airport grew from 29 million in 2002 to 36.7 million in 2007 (Ministry of Transport, 2008). Following the opening of Terminal 3 in the Changi International Airport in January 2008, passenger traffic continued to soar, recording a 5.4% year-on-year growth to hit 18.7 million passengers after operating only for half a year (Singapore Tourism Board, 2008).

While it is acknowledged that the Changi International Airport has performed remarkably well since its opening, it continues to face strong competition from other airports in the region as well as globally. In 2007, for example, the Kuala Lumpur International Airport in Malaysia recorded a growth of nearly 10% in passenger traffic, involving over 26 million passengers and creating a huge demand of over 7 million passengers who used the Low-Cost Carrier Terminal. Likewise, the Bangkok International Airport in Thailand had 42 million passengers in its first entire year of operations which began in October 2006. Similarly, Hong Kong also attained a growth of 7.5% in air passenger traffic, with about 48 million passengers in 2007 (Ministry of Transport, 2008). Relative to the Changi International Airport in Singapore, these regional airports have strengthened and are continuing to strengthen their capabilities to draw more passengers and attract more airlines.

Beyond the Southeast and East Asian region, there is also intense competition from airports in the Middle East. In April 2008, the Dubai International Airport overtook Singapore's Changi International Airport in terms of total passenger numbers in the first quarter of that year to become the sixth largest aviation hub in the world after Hong Kong, Frankfurt, Amsterdam, Paris and London. The Dubai International Airport achieved exceptional growth in passenger volume with an approximately 20% increase over 2007 (MCOT English News, 2008).

Besides intense competition from other airports worldwide, the Changi International Airport also faces other risks and challenges. This includes high fuel prices which can adversely impact the revenues of all airline businesses. In this context, the global financial crisis in 2008 has caused consumers to monitor their expenditures more carefully and to cut back on discretionary travel. With this backdrop, the Changi International Airport must continue to stay focused to deal with these challenges. Externally, any change in the global economic environment should be monitored closely. The airport management has to be pro-active to search for new opportunities and to prepare for threats that may arise. Internally, the airport management should pay close attention to the service quality the airport provides and to innovate relentlessly in order to stay ahead of the competition (Singapore Tourism Board, 2008).

With this agenda in mind, it is noted that the JIT concept has been implemented successfully in the manufacturing industry and the construction industry. However, past studies have focused on the operations and processes which occurred during the manufacturing or construction stage (Cheng and Podolsky, 1996). Various studies have shown that the JIT principles can be implemented successfully at the production stage (Lim and Low, 1992; Low and Chan, 1996, 1997). By moving upstream, one potential area where the JIT concept can likewise be implemented relates to site layout and spatial planning before production actually takes place. In this context, the JIT concept has been examined for good site layout planning for a construction site (Low and Mok, 1999) and for effective facilities design in ramp-up light factories (Low and Show, 2008). These studies suggest that the JIT principles are universal enough to be adopted in upstream space planning as well as downstream production processes. By extension, there is scope for considering the JIT concept in airport spatial planning and management for customers to experience enhanced service quality and efficiency.

If the JIT concept can be incorporated successfully into the design and spatial layout of airport terminals, this can provide a useful template for the airport management to continually consider and improve on the business environment to achieve enhanced productivity, service quality and cost savings. To accomplish this objective, a case study airport will be presented and the areas for potential improvement highlighted with JIT related recommendations proposed as appropriate. The case study chosen is the Changi International Airport in Singapore. The study will focus on improvements in the spatial layout in the Changi International Airport with a view to achieving the benefits that may be derived from the JIT concept. With these further improvements, it is hoped that the Changi International Airport will continue to maintain its position as one of the best airports in the world in the midst of growing intense competition from other airports elsewhere in the world.

For this purpose, the existing business processes and operations in both Terminal 2 and Terminal 3 in the Changi International Airport will be evaluated against the backdrop of JIT principles. There are three reasons for choosing the Changi International Airport for a start to benefit from this JIT-based evaluation and recommendations.

Firstly, the Changi International Airport has won many awards for being one of the best airports in the world. The accolades given to the Changi International Airport are well-recognized internationally. For example, in the first six months of 2009 alone, the Changi International Airport has won 12 out of the 14 best airport awards given out globally (Foo, 2009). Such global recognition and prestige received over many years suggests that the Changi International Airport has already put in place numerous best practices. These include, among many others, operational efficiencies in business processes relating to the checking-in facilities, passport control and security checking procedures.

Secondly, the Changi International Airport has made many great strides with respect to improvements over the years to meet the needs of its users. These include the noteworthy addition of Terminal 3 which started operations in 2008. The planning for Terminal 3 would have eradicated the perceived problems or weaknesses that were not resolved in Terminal 2 for various reasons of operational constraints. In addition, the design and development of Terminal 3 should also have considered as well as benefited from the lessons learned from both Terminal 1 and Terminal 2. In this context, it is therefore useful to compare between terminals to evaluate the enhanced changes and improvements designed and built into Terminal 3.

Thirdly, there is a lacuna on advancing the JIT principles for the planning and facilities design stage in airport development projects. Hence, the Changi International Airport offers an opportunity to evaluate how operational and business processes in a first-rate airport can be further enhanced through the relevant JIT principles. As Low (2009) and Liang (2009) noted, the Changi International Airport may score top marks in the world rankings, but there is always room for improvement. Likewise, Kaur (2009, 2009a) noted that despite its past successes, the airport should not rest on its laurels but should instead stay vigilant and be prepared to stay ahead. With the Changi International Airport being able to deliver exceptionally high efficiency and excellent service quality to its customers, this evaluation will allow other airports to consider adopting the JIT principles, albeit with the appropriate modifications made, in facilities planning as well as in spatial layout and design.

Beyond the aviation industry, this evaluation can also provide a model for other service-related industries and their associated supporting facilities to assimilate. Some of these other facilities can be found in integrated resorts, hotels and convention centers where service management and business efficiency are important consideration, just like in the Changi International Airport.

## Scope of Evaluation

The aim of this evaluation is, as a start, to examine how the business environment in the Changi International Airport may potentially be improved using the JIT principles for the airport to continually remain viable and competitive. To facilitate this, the airport management needs to ensure that the planning and facilities design to deliver the operations and processes both meet and exceed the expectations of the airport users. Three objectives will need to be achieved to meet this aim.

The first objective is to examine how the JIT principles can be applied to further develop and/or enhance the design of the airport terminals. Before this can be carried out, some modifications of the definitions of the JIT principles have to be made to apply these principles to the airport context. The same set of definitions for the JIT principles can then be useful for application in the future to the spatial layouts of other airports since these airports experience the same procedures covering check-in, immigration and other related cross-border travel operations.

The second objective is to review the application of the JIT principles to each airport feature. The application of each of these principles can serve as indicators of whether the airport features can bring about further enhanced benefits or have already achieved what they have been intended for the airport users in the first instance. The purpose of the review is also to highlight the airport features with existing low applicability where the JIT principles are concerned so that opportunities can be identified for these features to be further improved to achieve greater success.

The third objective is to propose recommendations for the appropriate JIT principles to be adopted, especially with respect to those airport features with currently low applicability. Airport planners and managers can then consider taking these recommendations into account to plan, develop and construct future airports or for the rejuvenation and re-development of existing airport terminals.

Due to the complexity of the subject matter and the limited access to the air-field and runways for obvious safety and security reasons, only the airport terminals will be targeted for the evaluation. The domains to be covered will include the following areas in the passenger flow chart as shown in Figure 1.2. The evaluation will take place in a progressive manner: commencing at the departure curb side, check-in hall, passport control area, transit mall, gate lounges, gate holding rooms, baggage claim area, the arrival hall and others. These areas are covered to ensure that the operations and facilities which the passengers use at each of these service points are efficiently planned and managed to provide excellent service quality.

However, the Baggage Handling System in the airport will not be evaluated more comprehensively than the other areas because of the level of complexity of the system as well as for security reasons. It is also noted that first-class passengers and very important persons (VIPs) take another pathway in Terminal 3 in the Changi International Airport. This arrangement is for privacy, efficiency and security purposes. This aspect of the airport operations will also be excluded from the evaluation because it concerns accessibility, safety and security issues. In summary, other than the two exclusions mentioned above, the evaluation will essentially cover the entire passenger routes taken by both the departing and arriving passengers in the airport.

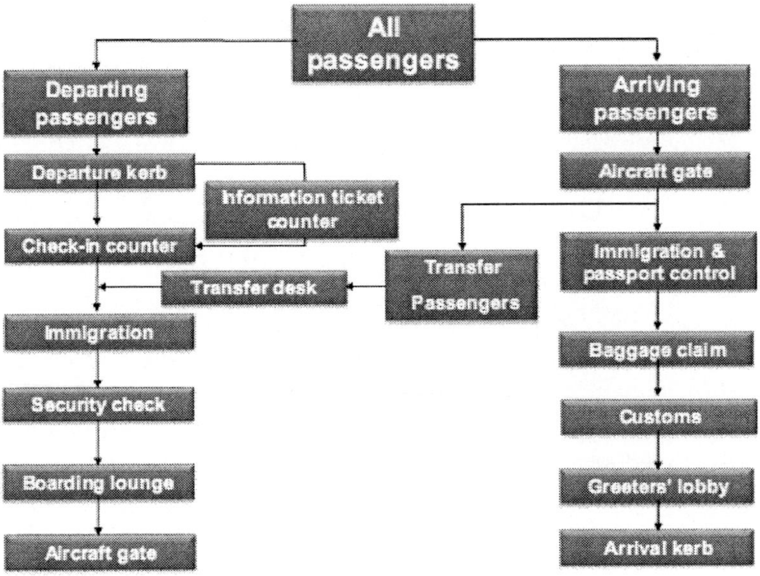


Figure 1.2. Passenger flows in airport terminals.

As most people would have visited airport terminals at some point in time, focusing on the airport terminals is therefore a logical choice to provide a better understanding of how airports operate and are managed within the context of the airport terminals. Since the JIT concept is the major focus in the evaluation, all the JIT principles will be discussed shortly with appropriate modifications made for interpretation and adoption within the physical confines of the airport. In this context, the main focus is to evaluate the application of the JIT principles in Terminal 3 of the Changi International Airport in Singapore. This is because Terminal 3 is also the latest addition to the airport and should have incorporated lessons learned from Terminal 1 and Terminal 2 that were built earlier in the airport.

However, among all the JIT principles, only the principle "continuous improvement" requires a basis for comparison. Historically, Terminal 1 of the Changi International Airport in Singapore was built with the purpose of handling large masses of passengers with more aircrafts and space. Its main planning concern in the early days then was on functionality to accommodate the increased demand in passenger traffic. On the other hand, Terminal 2 and Terminal 3 of the Changi International Airport were more similar because their planning had taken into account of customer service quality, operational productivity as well as efficiency. Hence, Terminal 2 was chosen for comparison with Terminal 3 in the evaluation. After narrowing down the scope of evaluation and explaining the rationale for the scope, the JIT concept will be further explained in the next chapter.

*Chapter 2*

# JUST-IN-TIME PRINCIPLES

## HISTORY OF JUST-IN-TIME (JIT) CONCEPT

It is a well-known fact that the JIT concept had been implemented positively in many organizations across different industries. Hence, it is worthwhile to examine the factors in the JIT concept that have contributed to enhanced productivity and quality in these organizations. This can be achieved through knowledge about the JIT background, the concept, characteristics and principlcs as well as to make appropriate modifications for evaluation in the present airport context.

It was said that elements of the JIT philosophy were first used in the Ford Motor Company by Mr. Henry Ford but only made its mark prominently when Mr. Taiichi Ohno, then the Executive Vice-President of the Toyota Motor Corporation, adopted it successfully as part of the Toyota Production System (TPS) in the early 1950s (Ansari and Modarress, 1990). The JIT concept was thereafter widely recognized following its phenomenal implementation in the Toyota Motor Corporation which transformed the corporation into a successful organization. The TPS is often also referred to as Lean Production because it emphasizes the less use of everything predominantly in the context of mass production; using half the human effort in the factory, half the manufacturing space, half the investments in tools, half the engineering hours to develop a new product in half the time and so on (Balakrishnan, 2003).

Although some elements of the JIT philosophy were observed to have been implemented in the United States historically, it was a well-known fact that Japan eventually became the first country to formally implement the JIT concept successfully through the Toyota Motor Corporation. It appears that the JIT concept was successful in the Toyota Motor Corporation because of severe constraints in Japan relating to limited land and space, as well as lack of natural and energy resources. Faced with such constraints, the Japanese strived to minimize and do away with any activities and operations that may lead to resource wastages. Other studies also suggested that the Japanese was successful here because they exercised good manufacturing management practices, operational discipline, consistency and teamwork (Lim and Low, 1992). Hence, the successful implementation of the JIT concept may have actually come from excellently planned management actions executed by the manufacturers rather than from cultural or environmental factors (Schmenner, 1990).

Apart from implementation in the manufacturing industry, other studies have also documented the introduction and application of JIT principles in the construction industry. The JIT concept has been shown to help construction firms to increase their productivity in various aspects (Low and Chan, 1996; Low and Mok, 1999; Low and Wu, 2005). However, one little known aspect relates to an understanding of how the JIT concept can be used to evaluate the efficiency of a completed construction site layout (Low and Mok, 1999) as well as that of a completed industrial building (Low and Show, 2008). This little known aspect of the JIT concept will be emphasized in this study for airports.

## JIT Concept

The JIT concept is a manufacturing philosophy based on the planned elimination of all wastes as well as continuously seeks to improve productivity (APICS, 2008). Much has been written about the JIT concept. One approach views the JIT concept as a philosophical way of organizational life that promotes an aggressive search for wastes as well as inefficient and ineffective activities (Thomas, 1987). This relates to one important JIT principle, elimination of waste, whereby any activity that leads to wastes should be identified and removed.

Another approach views the JIT concept as a philosophy that takes total pride in making the business more lean, simpler and more effective to operate with a higher degree of integration (Mortimer, 1986). Making the business leaner would require simplification and integration which may entail overlapping operations that will lead to greater efficiency. From a systemic perspective, the JIT inventory system which is an important part of the JIT concept is about having the right items at the right time, at the right place in the right amount and right quality without keeping inventory (Cheng and Podolsky, 1996).

The JIT concept is different from the traditional manufacturing process which is based on the Just-In-Case (JIC) workflow. In the JIC workflow, systems, materials and parts are pushed from one process to the next regardless of whether they are needed by the next station (Low and Mok, 1999). By doing so, the purpose of the JIC workflow is to reap the benefits of economies of scale from intensive production. However, the JIC workflow has little concern over the actual production needs. Eventually, there would be stocking up of inventories which not only take up space but would also incur costs from storage, redundancy, spillages and maintenance. In addition, and more importantly, products of inferior quality, defective components or even rejects might even go undiscovered as these would be hidden and concealed among the stockpiles of inventories.

How is the JIT concept then different from the JIC workflow? In this context, the JIT concept will only produce just in time for the subsequent process. If the next process does not need the current work/parts at such a fast pace, the workers will stop or adjust their speed and work accordingly to synchronize with and meet the needs of the next work station. If there is no need for the end products or parts, the production will stop and the workers concerned may be deployed to other work stations to help other workers. The ability to multi-task is therefore an important function in the JIT concept.

## JIT Concept in Organizations

The JIT concept has been proven time and again to be successful for firms based in Japan. However, there was minimal publicity of this concept outside of Japan until Kawasaki in Lincoln, Nebraska in the United States successfully adopted the concept (Schroeder, 1989). Following the Kawasaki's experience in the United States, other firms such as Boeing, Southwest Airlines, General Electric, Hewlett-Packard and IBM have also since then took into account the JIT concept in their manufacturing processes (Low and Chan, 1997).

In addition to these firms, the automobile industry is and has become an ideal example of how the JIT concept may help to improve the efficiency of the entire production process (Karlsson and Norr, 1994). Faced with the large number of complicated manufacturing processes in the automobile industry, management must find ways to raise productivity through integrating work processes and reducing unnecessary wastages. In this connection, the JIT concept has helped to boost the production capacities of automobile manufacturers for them to be responsive to market demands. The JIT concept has helped automobile manufacturers achieved higher profits and competitive edge over others.

Goddard (1986) suggested that companies can achieve the competitive edge by competing on the basis of cost, service and quality. With the JIT concept, companies can eliminate waste during production to save cost and yet remain reliable in meeting customer needs with quality works. There are three main manufacturing objectives for using the JIT concept (Shingo, 1982). Achieving these objectives will help add enormous value to organizations.

The first objective is to increase the organization's ability to compete with rival firms so as to remain competitive in the long term. Organizational competitiveness is enhanced through the use of the JIT concept because organizations would strive to develop optimal processes for their manufacturing products (Cheng and Podolsky, 1996). The optimal process would lead to the next goal of increasing the degree of efficiency within the production process. High efficiency will lead to productivity and thus, reduction in costs of production.

Besides increasing productivity, quality is also enhanced. With quality works, there will be no need for re-works or performance checks since defects are found and eliminated at source. Another characteristic that contributes towards quality is the value-add to products or services. This only involves conducting activities and processes that add value to products which build in product quality (Cheng and Podolsky, 1996). For any organization, quality is an important aspect in attracting customers to keep coming back for the products or services. With these regular and repeat customers, there would be referrals to others which will in turn bring in more customers. In this way, firms can continue to prosper, leading to further business expansion.

The last objective is to reduce the levels of wasted materials, time and efforts involved in the production process. This will cut down on costs and hence, increase profits. Thus, firms who are willing to take up and implement innovative ideas on the JIT concept for their operating processes will gain a competitive edge over those who are unwilling to do so. These forward-looking organizations can then go on to vie for greater success with other firms in the long run.

## ADVANTAGES AND DISADVANTAGES OF JIT

The many advantages associated with the JIT concept when this is carried out effectively as described above include improved service rendered for customers, more productive operations, better work environment and lesser costs incurred. In doing so, other potential benefits follow. There will be higher employee morale, motivation and teamwork. Teamwork is cultivated since employees have to communicate more often in order to integrate processes and other related tasks. Other benefits may include more reliable working relations with suppliers and improved customer satisfaction (Cheng and Podolsky, 1996). Timely delivery of parts and services helps to build trust in the working relationship. Looking at the overall picture, there are increased financial benefits which is the main reason that drives most organizations to adopt the JIT concept.

Despite having numerous advantages, there are disadvantages associated with the JIT concept as well. A fundamental weakness of the JIT concept is the high interdependency in all linked processes. This whole process is vulnerable to how employees or suppliers work. Any delays in supplying the parts or other services can cause disruption to the entire process or supply chain. Consequently, inventories may originally be kept as buffer stocks. There will be no back-ups for sudden, unexpected circumstances when adopting the JIT concept. Hence, once a problem is left undetected, it can snowball into major issues which can be difficult to solve thereafter and which may disrupt the whole working process.

The scenarios described above can have serious implications on the firms involved. Using the JIT concept can appear to be risky to certain businesses. Such businesses may become vulnerable in the supply chain arising from situations caused by labor strikes, interrupted supply lines, market demand fluctuations, stock outs, lack of communication upstream and downstream in the supply chain and unforeseen production interruptions (Greenberg, 2002).

The weaknesses inherent in the implementation of the JIT concept must therefore be thought through carefully. In this context, Johnson (2001) noted that some companies have found that their just-in-time manufacturing systems have let them down. Companies need to critically evaluate the advantages and disadvantages before incorporating the JIT concept within their production systems. Although the JIT concept may appear to exhibit weaknesses, however, in most cases, the benefits seem to far outweigh the risks for the JIT enabled company (Broyles, et al., 2005). The next section will examine how some of these disadvantages can be turned into successes for the JIT concept to work.

## SUCCESSFUL JIT INGREDIENTS FOR ORGANIZATIONS

It is vital that companies prepare and implement the JIT concept in line with their business plans. Because each company has its own strategic and organizational goals, certain JIT principles may be more critical and may therefore take precedence over others. Alternatively, companies may re-evaluate the initial intended business plans to examine how these may be tweaked to fit suitably with the more vital JIT principles in order to succeed.

After this step, the JIT concept would require the cooperation of the entire organization and suppliers to make the JIT concept work. Top management has to be convinced first on the workability and feasibility of the JIT concept. Only then can the organization get the commitment from other employees. It would require the effort of the entire organization to execute the new plans together and collectively in order to do well. However, it should be noted that it may not be that easy to reap the success of effective implementation in this area because of several obstacles. More significantly all humans tend to resist change, be it physically or mentally. Hence, the implementation plan needs to initiate ways to improve worker commitment, to instill a sense of responsibility and to increase work satisfaction.

## SUCCESSFUL JIT INGREDIENTS FOR OPERATORS

The JIT concept requires strong reliability and consistency of all links in the supply chain for it to work (Cheng and Podolsky, 1996). Trust and stability need to be built up over time and must be maintained. A flexible spatial layout and an open-minded manpower resource are highly desirable because the JIT concept responds according to the needs demanded on the spot. Identifying and solving problems at source is also required of workers across all areas in the supply chain to prevent these problems from degenerating into serious problems with dire consequences. This issue will be discussed further in the section on the JIT principles.

## SUCCESSFUL JIT INGREDIENTS FOR END USERS

In every business, the most important focus is on how the end users feel about the final products or services experienced. Since their opinions are highly valued, customer involvement is necessary in product design, development and delivery (Bowen and Youngdahl, 1998). Hence, if possible, customers should be contacted early for their views for improvements in the planning stage in order to achieve good quality and ultimate satisfaction.

For the JIT concept to work well, it can be seen that its implementation requires combined efforts from all the above three stakeholders (i.e. organizations, operating employees and end users). Production can still proceed without any one of these three stakeholders, but the resultant effect is that quality and strategic design development might be compromised to some extent. Ultimately, success is highly dependent on organizations to start off and initiate the necessary actions in implementing the JIT concept in their operations.

## JIT PRINCIPLES

Therc are seven principles under the umbrella of the JIT concept. These are:

1. Elimination of Waste
2. Kanban/Pull System
3. Uninterrupted Work Flow
4. Total Quality Control
5. Top Management Commitment and Employee Involvement
6. Supplier Relations, and
7. Continuous Improvement.

The key provisions in each of these principles are discussed below.

### Principle 1. Elimination of Waste

Elimination of waste in manufacturing defines waste as anything other than the minimum amount of equipment, materials, parts and working time absolutely essential to production (Hay, 1988). These essential resources refer to machinery, materials and manpower that increase or add value to the product as they are used to produce the product. On the other hand, actions such as shifting, storing and inspecting, do not increase or add value to the product. There is therefore no value gained from doing these actions in the processes leading to the final products.

There are seven major causes of waste as identified by Shingo (1982). These are:

1. Waste from over-production
2. Waste of waiting time
3. Transportation waste
4. Processing waste
5. Inventory waste
6. Waste of motion, and
7. Waste from product defects.

The nature of these wastes is discussed below.

***Waste from Over-Production***

Over-production waste simply refers to producing more than is necessary. This is often in response to the fear of producing too little for consumption and running into shortages. This causes inventory level to increase when there should be none if the exact quantity demanded is produced. As a result, cost will also increase to pay for the space needed to store the additional inventory and the manpower for handling the movement of these inventories. One way to solve this problem is to educate people not to produce excess inventory. To help them do so, they should be encouraged to present level schedules so that workers are able to keep track of how much to produce, no more and no less.

***Waste of Waiting Time***

This is the most common type of waste identified as there are many sources that can cause this waste. One example of such sources can come from equipment set-up or start-up time. Another example is the time that operators wait for machines to warm up. Despite waste coming from many sources, unnecessary waiting time can actually be prevented through careful planning and taking corrective actions if wastage in waiting time is discovered in certain production processes.

### *Transportation Waste*

Transportation waste is caused when an item, which is being moved, is stored temporarily or being re-arranged unnecessarily (Lim and Low, 1992). For example, transportation waste commonly occurs when a particular part in a production process is transferred from the machines across the factory when it could have been located together with the other parts along the production line. One such case is the casing and battery of a mobile phone. The machine for placing the casing to the phone is placed before the machine for placing the battery. After inserting the battery to the phone, the process has to move backward to finish the product with the casing before moving forward for packaging. This results in unnecessary movements that lead to inefficiencies. To solve such problems, there is a need to improve the layout or coordination of processes in order to minimize transportation waste.

### *Processing Waste*

The nature of this waste suggests that an item is going through more processing efforts or longer processing time than required. This can arise from poor maintenance or late preparation of machines or equipment which requires additional time and effort to start up or to make them perform up to their optimum standards. Operators may need to check on the machines regularly to prevent such wastefulness or stoppage from happening. Some helpful actions can be used to improve the situation, for example, by applying lubricants to the machine parts to achieve a smooth process.

### *Inventory Waste*

Inventory waste refers to storing spare inventory which usually results in incurring higher costs, more time and extra space to handle, although the purpose is to avoid shortages. Some ways to eliminate inventory waste can include producing only items that would be needed in the subsequent station or to supply only the quantities needed for production even though some parts may be purchased in bulks.

### *Waste of Motion*

This waste of motion (or effort) occurs when an item is shifted to another place needlessly or when any motion which adds no benefit to the product takes place. For example, the workers are required to walk a distance to get a part for production and then to walk back to continue with the production. Simply by just placing parts and tools close to the workers or improvising on the arrangement of items can help to eradicate waste of motion.

***Waste from Product Defects***

This can happen if quality control is not exercised by the workers to ensure that only defect-free products are delivered. Arising from these defects are re-works where extra resources, time and effort have to be expended to produce another one that is defect free or to carry out repair works. Ultimately, this leads to increased cost and delays in the final production. One good method is to recommend employees to take instant corrective measures once defects are found. This is to solve problems and rectify defects at source rather than to allow these to fester along the production line. Getting rid of this waste generates benefits such as increased productivity, reduced space requirements and shortened production time.

## Principle 2. Kanban/Pull System

There are basically two ways to manage the production process. One way is the push system whereby production, based on forecasts, is made with no consideration to the actual demand for the products or component parts and where these products are "pushed" out to the customers regardless of whether they want the products or not. Another way is the pull system, also known as the kanban system, which only produces when there is actual demand for the products or component parts. The word "kanban" means "visual record" in Japanese. The kanban system generally refers to a form of visual mechanism used by workstations to signal the need for more parts or components from the preceding station. In short, the kanban system generates parts/products just in time and in the right quantities when required, no more and no less.

However, because the kanban system does not deploy any backup plans and the volume of quantities is produced just enough to meet demand, stoppage of production cannot afford to happen. To implement this JIT principle, the company involved needs to set up a kanban mechanism so that problems such as inventory or bottleneck issues from a machine can be spotted and be rectified instantly. In addition, at the early stage of implementation, the company may face short term fall in efficiency for the workers to familiarize themselves with the new coordination methods. However, the company will be able to save costs in the long run resulting from the elimination of inventory.

One of the reasons why the kanban system is important to airport management is because of the benefits that the system can bring about; namely improving quality standards, fulfilling customer needs and enabling airport operations to achieve smooth work flow.

## Principle 3. Uninterrupted Work Flow

The JIT manufacturing model demands an uninterrupted work process (Low and Show, 2008). One way to achieve this is by restructuring the layout so that similar or related items or processes can be placed together, one after another. The purpose is to minimize movement and handling of these items or processes with minimal disruptions (Low and Chan, 1997). Another way is by simplifying the work processes by doing the job correctly at the very first time so that reworks or rectifications will not be needed later which interrupt the smooth work process.

To accomplish uninterrupted work flows, this JIT principle calls for the following four improvements to be made: namely the usage of "focused factory" and "group technology" (GT) techniques, design for simplification and automation, reduction of process set-up times and implementation of machine/equipment Total Preventive Maintenance (TPM) programs (Low and Chan, 1997).

For the first improvement on focused factory and GT, both techniques are management philosophies which identify and exploit the "sameness" of items and processes used in manufacturing (Skinner, 1974). Focused factory refers to the production of a single family of parts arranged in a manufacturing layout. GT is about equipment arrangement required in producing a family of similar parts step by step accordingly, thereby connecting all associated operations in the process.

In designing for simplification and automation, things must be done in easy steps by doing the job correctly in the first attempt. Only when things become simplified, then automation can be adopted. For example, to produce a car, there are many parts and processes involved. Next, these processes must be broken down into simple steps to be carried out. When finally broken down into simple steps, automation can be incorporated to facilitate these repetitive steps to produce the final product-car efficiently.

Reduction of process set-up times refers to the time needed for a machine to be set up to conduct a new process. Doing away with this set-up time waste will increase efficiency and flexibility. This should be considered together with the implementation of machine/equipment Total Preventive Maintenance (TPM) programs as well. When a machine breaks down, it will cause disruption to the process and this will in turn, give rise to poor service quality. This also shows that the machine is unreliable but it can be prevented with the set-up of a TPM program. The aim of the program is to be proactive to prevent breakdowns from happening rather than to spend time and money for repairs when these happen. Hence, this accounts for the reason why maintenance is best left to the care of the operators as far as possible as they have the responsibility to make sure the machines work well all the time. In addition, they are in better control of the machines at the operational level and have better knowledge to detect the signs of impending breakdowns compared to other employees.

## Principle 4. Total Quality Control (TQC)

The TQC goal is to get the job right the first time, all the time. To achieve zero buffer stock, the kanban system is put in place to attain a smooth and continuous process from one work station to another. For this to happen, the components of each work station must be completed with quality and with perfection before handling over to the subsequent work station. When a component is found to have quality problems, the operational process involved will need to be stopped or put aside for checking. Therefore, in order to achieve a good quality and smooth process, every worker is personally responsible for the final product manufactured. If everyone makes the effort to ensure quality, the act of inspection can be eliminated and no extra cost will be incurred for re-works.

However, the "quality" in TQC does not simply refer to quality as a feature of the product but to the business processes (Cheng and Podolsky, 1996). The term "quality" can be divided further into three types: namely quality of design, quality of conformance and quality of service. Quality of design refers to the design of goods/services which meets or exceeds customer expectations and is simple to manufacture. Quality of conformance would be to achieve the design specifications in the final output whereas quality of service would refer to quality in all dealings with the customers when doing business. These can be measured through various parameters like timing, technical performance, comfort and reliability (Cheng and Podolsky, 1996).

There are several advantages to having good quality control. There will be cost reductions resulting from improvements, minimal quality problems and for production to proceed normally. In addition, in the JIT manufacturing system, faults or errors related to some quality problems probably mean that some sections of the system may have gone wrong. It is thus a good chance to learn from these faults for further enhancement and development (Low and Chan, 1997). Hence, having a total quality control system is an important factor that contributes towards the successful implementation of the JIT manufacturing system.

## Principle 5. Top Management Commitment and Employee Involvement

This JIT principle is highly dependent on the responsibility and commitment of every employee in putting in effort to apply the appropriate JIT principles, especially in areas relating to elimination of waste and continuous improvement. Employees need to change their mind sets in the way they work. This involvement includes not only the top management, but also all employees throughout the entire organization. This is important because the top management is only primarily involved in planning and other management functions but may not execute the work *per se*. That is why top management may be unable to understand the actual working process and cannot implement further improvements due to their lack of actual work experience on the ground. Hence, only the workers directly involved can provide better insights to the improvements that can be made. But first of all, top management has to give the workers the chance to make or suggest the changes. The result of this active participation by all employees will bring about higher productivity and quality improvements in the production processes.

To implement this principle successfully, the commitment and cooperation displayed by the top management is highly valued. If the top management is supportive, the middle and lower management is likely to follow suit. Ultimately, this will set all employees to work towards the same goals of making continuous improvements to bring about better service quality than before. Thus, efforts must initially be made from the top management to introduce the new systems i.e. suggestions to encourage the lower management personnel to join in the discussions during the inception period in the early exploratory stages of an initiative. When employees are given the opportunities to make decisions, they are instilled with the responsibility and motivation to do their part for the well-being of the company.

In addition, the JIT concept sees skills and training as important for all employees. It is concerned with whether workers are given broad-based training to perform a wide variety of jobs. This will in turn provide the companies with flexibility in their work performance. Furthermore, properly trained employees may be able to come up with creative ways to improvise on operational methods in the long run, especially those in direct contact with the operations on the ground.

Having a good habit of keeping the workplace organized and neat with regular checks on the machinery is also necessary. This is to be accomplished in line with the TPM program. Employees are to play an important part in ensuring that the quality of work produced at the end of the day is flawless. In addition, problems can be identified more readily in a tidy and clean work environment than one that is littered in a messy environment. This requires the cooperation of all employees to have a clean, neat and organized work environment.

## Principle 6. Supplier Relations

Besides the involvement of employees, there is also the need to have good relationships with the suppliers to satisfy other principles like the kanban system and to achieve continuous improvement. This good relationship will then lead to long-term partnership which results in stabilized operations that can bring more certainty to the work environment. It is preferable to stick to one reliable supplier or just a few sources of supplies instead of switching frequently between suppliers. This arrangement can lead to improvements in communications, consistent quality and business certainty in the long term.

For production, the quality of a completed product relies heavily on the quality of the materials and components needed to finish the product. To achieve good quality control, the choice of suppliers is one of the vital ingredients for success. The special feature of suppliers in the JIT concept is that they are required to make deliveries in the exact quantities and frequencies demanded at the times when needed for production.

## Principle 7. Continuous Improvement

Continuous improvement is what the JIT concept constantly seeks to achieve. After the JIT concept is put into operation/service, there should be improvements shown. If the actual results differ from the expected performance, these performance gaps shall be further evaluated to uncover the deficient areas that will be analyzed further as areas for improvements. Hence, the JIT concept is an endless process which sets in motion the mechanisms to improve continuously in the long run (Low and Chan, 1997).

With this principle, organizations will be instilled with the attitude to improve things all the time. This ties in well with the earlier employee involvement principle whereby employees of all levels need to share their experiences and observations at work to solve problems or discuss potential improvement areas (Hay, 1988).

## RELATIONSHIP BETWEEN JIT PRINCIPLES

Having reviewed and explained the seven JIT principles, it can be noticed that each principle does not just exist on its own. At times, one principle can support, enhance, complement or reinforce another principle. For instance, there can be elimination of waste in an area. This can also constitute as a form of continuous improvement since it helps to reduce wastage. Hence, it should be recognized that there can be overlaps in the linkages between the JIT principles. In the final analysis, the JIT principles should not be viewed upon as individual, stand-alone principles. Synergy can be obtained by appropriately implementing all JIT principles at the same time in a single production scenario.

### JIT Principles in Airport Management

One apparent difference between the manufacturing sector and the airport context is the complexity of airport operations in the latter. In most major airports, the working procedures are more uncertain due to the complex flow of users at different times of the day. This is in sharp contrast with the standardized working procedures found in most factories. Although there are standard procedures for passengers to follow in the airport, how the operations are carried out would depend on the passengers themselves. Passengers cannot be controlled in terms of their preferences, priorities, walking speed as well as their baggage and check-in methods. This is quite unlike the manufacturing sector where companies have stringent control over workers' efficiency and quality management systems.

Conversely, excellent service quality is of utmost importance to airport management. This means some of the JIT principles have to be modified appropriately to cater to the way in which the airport works. One such instance is that there is always a constant need to keep inventories such as the trolleys for travelers to put their baggage. There appears to be no definitive way to anticipate the exact number of trolleys that passengers will use at any one time. One solution is for the airport management to provide for more trolleys rather than less to maintain a good level of airport service. Besides keeping inventories, there are many other areas that the JIT concept can help to provide significant contributions for the airport to achieve excellent service quality. The JIT concept will therefore be evaluated within the context of the fundamental problems that may impede the progress of an airport in terms of efficiency, productivity and profitability. This will also provide a first account of how the JIT concept can be applied to assess the degree of service quality in airport operations as well as provide a platform for airport management to reflect and strive towards continuous improvement.

*Chapter 3*

# CHANGI INTERNATIONAL AIRPORT

## INTRODUCTION

The objective of modifying the JIT principles for airport management will be incomplete if only the JIT's side of the story is examined. To provide a more complete picture, similar studies are needed to provide a better perspective of how the JIT principles can be relevant in the airport operating environment. For this purpose, the Changi International Airport in Singapore, being one of the leading and most advanced major airports in the world was identified as a suitable airport for further evaluation of the airport operating environment.

## BACKGROUND OF AIRPORT

The Changi International Airport has a long history dating back to when Terminal 1 first commenced operations. There are currently three terminals operating in the Changi International Airport. Since its opening in 1981, the airport has made its mark in the aviation industry as a benchmark for service quality excellence, winning over 340 awards since its opening and with 19 Best Airport awards won in 2007 alone (Changi Airport, 2008). At the time of this study, the international accolades for Changi International Airport continued to flow in worldwide. Being recognized repeatedly as one of the "Best Airports" globally for its high efficiency, it is worthwhile to examine what Changi International Airport has done right strategically to improve continuously for it to maintain its world renowned reputation. This study therefore aims to examine the planning and operating processes of Terminal 2 and Terminal 3 in the Changi International Airport through the use of the JIT principles described in the earlier chapter.

Changi International Airport Terminal 2 started operations in 1990, followed by an extension project in 1996. Deploying a linear configuration that runs parallel to the runways, it is located adjacent to Terminal 1 towards the south. Terminal 2 also heralded in the opening of the original sky-train system that links the two terminals via the landside (Flight Memory, 2008). It occupies an area of 358,000 square meters and is able to handle 23 million passengers every year. Terminal 1 was built with the planning concept of functionality. As aviation activities increased over the years, planning concerns were shifted towards providing passengers with enhanced quality in terminal design and improvements in ambience together with technological advancements.

Changi International Airport Terminal 3 was opened in 2008. The S$1.75 billion terminal, comprising of a space area of 380,000 square meters, is located directly opposite Terminal 2. It is the largest among the three terminals. The terminal itself has seven floors with three levels of basements and four levels above ground. Various services and amenities such as necessities outlets, retail shops as well as food and beverage outlets are provided for employees, passengers and users. All in all, Terminal 3 focuses on four design principles that are underpinned by clarity, natural lighting, external views and maintainability to derive the best value for its users.

## GOOD AIRPORT DESIGN AND MANAGEMENT

Following the introduction to the three terminals in the Changi International Airport, some key concepts and principles relating to good airport design and management are described below.

Flying can be a naturally stressful experience. This is because of the queues, the waiting time involved, the crowd, the baggage and the anxiety faced in trying to catch the plane in time. An important aspect that passengers look out for in a good airport design is whether they can get onto their planes on time. The five key considerations for good airport design include the building design, signage, transport links, having minimal queues and retail (Geoghegan, 2008). These are discussed below.

## Building Design

A good building design is one that can reduce the anxiety level of the travelers (Geoghegan, 2008). In this context, it was noted that in some overseas airports, difficulties with way-finding occur the moment the travelers step into the terminals, which is undesirable because this lack of directional certainty can add to their anxiety level (Geoghegan, 2008). Hence, good terminal planning and design need to be made clear to any user, especially the travelers who may be stepping into the airport for the first time.

## Good Signage

As a basic requirement, airport signage needs to achieve the conditions of certainty, clarity and reliability. This means that when any airport user needs to look for certain things, the signage nearby will be able to provide the correct directions that will eventually lead the user to the final destination in the shortest possible time. Having too many signs in an airport can also be bad as these can be too distracting as well as confusing. More importantly, the signage should be made internationally the same for universal recognition. This will then allow the users to be familiar and comfortable with directions regardless of the origin airport or the destination airport that the users or passengers arrive at.

## Transport Links

The airport itself may be very nicely done up physically but the planners need to take note of the journey to and from the airport via the city centre. A fraught journey can make passengers feel irritated getting to the airport, especially if they are rushing to catch their flights. A good example to learn from is the transport links to the Hong Kong International Airport where its Airport Express train can take passengers from the downtown area to the airport in about 20 minutes.

## Queue

Although the processing of passengers cannot be controlled readily, a good design can help to minimize the queue time and the associated hassle. It is acknowledged that it is mandatory for all travelers to go through all the airport immigration and security procedures. However, while queuing, passengers' attention can be diverted to some extent by providing nice views in the vicinity, good acoustics and day lighting around them. If the overall ambience is good, the passengers will feel comfortable while waiting. As a result, time will also be perceived to pass by faster by the travelers. Other electronic means for immigration clearance in airports may also help to reduce or eliminate the need for queuing.

## Retail

Experts in retail architecture have noted that shopping is essential in the design of every airport (Geoghegan, 2008). Retail brings in a huge amount of revenue for the airport yearly. However, aside from generating income, the retail outlets and the shopping experience must be designed to provide passengers with excellent airport service instead of "herding" them to shop.

## Inadequate Airport Design

Quite apart from the good design features mentioned above, there are also poor designs which have left passengers with bad impressions while using the airports. One apparent bad design is when the signage for shops and restaurants is made more obvious than the perceptibly more important signs for boarding gates and passport control area. The rationale for this is that shops and restaurants may generate more income for the airport. However, this might neglect the group of frustrated passengers who are rushing to catch their flights. In addition, there might be some advertisements or promotion banners which distract travelers from the existing and more important signage that leads to the boarding gates and security control areas.

Besides signage, there are also other issues related to poor airport design. Taking an international airport for a major city in the United States as an example, postings in Skytrax (2009) have suggested that there were frequent instances of missed flights. This was because the airport is situated too far away from the city (about 100 miles) and the walk to the boarding gate is too long when compared with other international airports. This feedback appears to be reliable because Skytrax (2009) is one of the world's largest platforms of reviews for airlines and airports that cover over 620 airlines and 645 airports. There were also similar comments posted on a major international airport based in the United Kingdom. Although the people who posted these comments were usually the ones who have issues with the airports through their first-hand experience, these are indeed the pressing issues that should not be swept aside by the airport management.

## Design of Changi International Airport

As observed by experts in airport design, the function of an airport should drive the form (Geoghegan, 2008). In this context, some of the best-functioning airports have the simplest designs and that Singapore's Changi International Airport is one such instance with a fairly basic concrete building, accompanied by wide corridors, plants and greenery, as well as good lighting.

When the Civil Aviation Authority of Singapore (CAAS) started to plan for Terminal 3 at Changi, it concentrated on, among other things, the following four guiding principles: clarity, natural lighting, external views and maintainability. The aim was to ensure that travelers can simply navigate their way through the terminal with minimal dependency on signage (Singapore Sights, 2008). In addition, the see-through layout allows visitors to easily orientate themselves in relation to where they wish to go. The abundance of soft, natural lighting also creates a comfortable ambience within the building. The open plan design can be easily cleaned and maintained, with the associated technical and mechanical equipment hidden away from the public view. With all these features, Changi International Airport Terminal 3 strives to provide a comfortable, safe and enjoyable experience for its travelers (Lim, 2008).

The notable observation of Changi International Airport Terminal 3 is that while it is huge, yet it has a good building layout that is easy to get around with the aid of the signage, transit system and travelators. Nevertheless, it was noted that there seems to be a longer walk to the boarding gates in Terminal 3 when compared to the other terminals. Quite apart from this longer travel distance, it appears that the design concept for Changi International Airport Terminal 3 has all the hallmarks of an excellent international airport by global standards.

## Good Airport Management

Striving for customer service excellence should be a very high priority for airport management (Broadbent, 2008). To qualify as good airport management, customers primarily look for punctuality of flights, efficient check-in and immigration procedures and excellent airport facilities for the long-haul travelers. First of all, passengers loathe waiting and would find it irritating to have their flights delayed. Although the delays may be caused in part by the airline companies, the airport management should exercise some control over them to minimize their occurrence. This is because travelers will generally refrain from using an airport which often experienced flight delays for whatever reasons and no matter how good the airport clamored to be in other areas.

Secondly, passengers loathe queuing for too long. There are a few stations or check-points in most international airports which a majority of the ordinary travelers need to queue. These include stations for check-in, passport and immigration control, security checks and boarding gates. If the queue for all of these stations is very long, the passengers might choose not to use the airport and switch to another viable alternative in the future. In this context, it is worthwhile to consider airport facilities that will especially benefit the early check-in or transit passengers who will tend to stay longer in the airport than the final destination passengers who will tend to exit quickly from the airport. There should be a good variety of shops and restaurants for the former group of passengers to choose from while waiting in the airport. Service employees in the airport should also play a part in helping to enhance a pleasant airport experience. Employees working in these shops need to be trained to treat customers with the right work attitude and to exude a high level of professionalism and service excellence.

## Managing Changi International Airport

It appears that the management philosophy in the Changi International Airport adopts a holistic approach towards coordinating the interests of the relevant government agencies, airport concessionaires, tenants, suppliers and users, i.e. the passengers. This demands meticulous planning and execution in order to serve the customers well. In this context, it is worthwhile to note that there are various categories of users of the airport: including the originating passengers, destination passengers, transit and transfer passengers, well wishers, greeters and casual visitors who come to the airport for leisure purposes. Hence, it is important to cater to the diverse needs of each of the above categories of users, which the management of the Changi International Airport has admirably accomplished, as evidenced from the many international accolades which the airport has won consistently for many years (Changi Airport, 2006).

One of the main reasons why the Changi International Airport is continuously recognized as one of the best airports in the world relates to its high efficiency in carrying out its airport operations and procedures. Passengers can expect a speedy exit from the airport and fast baggage claims; likewise for check-in and passport control procedures (Changi Airport, 2008). In this context, it is highly unusual to find any adverse feedback commenting on any serious delay issues which might otherwise suggest that the Changi International Airport had slacked in this area. On the contrary, travelers have frequently commented positively that they liked the facilities provided, such as the free movies, free internet access, the good variety of retail shops, 24-hour supermarket and the reasonable prices charged (Skytrax, 2009).

While it appears that the Changi International Airport has already excelled in many ways where operational and procedural management is concerned, it is postulated that the airport should not rest on its laurels. Instead, like many great organizations and companies worldwide, the airport should consistently strive for continuous improvements in all its endeavors, big and small, to further enhance the customer experience (Balakrishnan, 2003). It is with this in mind that this study posits that continuous improvements for the Changi International Airport can come about by benchmarking its existing operations and procedures with the JIT principles that have earlier been presented.

*Chapter 4*

# RESEARCH DESIGN AND METHODOLOGY

## RESEARCH DESIGN

The case study approach forms the basis for the research design. A case study is an intensive study of a single group, incident, or community (Kumar, 2005). This research design fits the goal of reviewing the appropriate JIT principles in the Changi International Airport. Since the aim of this study is not to solve specific problems or to quantify items, the case study method is more suitable compared with the use of other methods such as experiments or surveys. Moreover, case studies are based on both quantitative and qualitative evidence which are the two avenues for uncovering the information for this study.

The main objective of this study is to evaluate how the JIT concept can contribute to improving the business environment of an international airport. Information pertaining to these issues will come from consultations with the relevant airport workforce and the airport users. However, for a start, this case study will only focus on the airport workforce who is familiar with the airport operations, procedures, management and planning. Hence, the target population for the research centers on those airport employees with the relevant knowledge of how the airport functions.

The sampling frame should be representative of the population workforce. This is because among those who participated in the research, some were involved in the planning for Terminal 2 and Terminal 3 in the Changi International Airport and would best understand the planning considerations as well as constraints. As the research focuses on the airport employees, the sampling method may not appear to be applicable in this case. Given the small pool of employees who have the related knowledge, the only viable choice is to secure the maximum possible number of airport employees to participate in the study. In this context, the minimum sample size was 30 respondents, made up of airport employees for the questionnaire survey.

## Research Methodology

Three methods were adopted for the empirical part of this study. These included the observational walk-through, interviews and survey questionnaires. The reasons for their use are discussed below.

### Observational Walk-Through

The observational walk-through method provided the research team with the opportunity for site visits to better understand how the airport operates as well as to observe first hand, how the JIT principles may be of relevance.

#### *Justification for Walk-Through*

As part of this research method, the site visits provided the research team with the opportunities to explore Changi International Airport to obtain the relevant information and to learn to use such information as background knowledge. In the process of familiarizing with the various operational aspects of the airport, the potential areas for further examination can then be identified. This aids in the decision making process for finalizing the research scope. Real life experience in a walk-through enables one to discover things that might not otherwise be found readily in the reading materials or media.

More importantly, the comprehensive site visits have helped to identify most of the JIT principles that can be applied in the airport context. This has been found to be extremely useful in identifying the key contents of the study. To reiterate, the study is based primarily on thinking through all the airport elements, for example, the airport security screening system, before identifying and analyzing the relevant JIT principles that the airport planners have adopted either directly or indirectly in the airport. The opinions of the airport employees relating to the importance, significance and applicability of the JIT principles were sought thereafter based on these findings.

***Conducting the Walk-Through***

Having reviewed the JIT concept and its principles, a template was formulated with a view to understanding how the seven JIT principles work in the airport context. The next stage was to gain more knowledge of the spatial characteristics of Terminal 2 and Terminal 3 in the Changi International Airport. This was accomplished through the relevant websites and press reports. Prior to the start of the study, three site visits were granted to the research team to understand the planning, operations and management of both Terminal 2 and Terminal 3 in the Changi International Airport. These site visits were granted following an application to the relevant airport authorities to observe how the airport was managed with a view to exploring areas for continuous improvements that may further benefit the airport in the near future. The application was granted following an explanation of the purpose of the study. The approval clearly reflects the forward-looking and pro-active stance of the airport authorities in wanting to identify areas for further enhancement.

The first visit was a walk-through in Terminal 2 and Terminal 3, led and accompanied by senior airport personnel in July 2008. The main purpose was to learn about the operations and the planning processes which transformed the two terminals to what they are today. The visit took about three and a half hours to complete. The areas covered included the departure halls, concession areas, passport control areas, transit malls and arrival halls of both terminals. For the gate holding rooms and the areas for arrival immigration, there was limited accessibility for viewing in Terminal 2, obviously for reason of security. The lessons that can be derived from this site visit will be further elaborated later.

In the second site visit, which was also conducted in July 2008, the Budget Terminal and its operations were examined. The Budget Terminal caters specifically to the budget airlines that land in Singapore. The systems, layout and planning processes in the Budget Terminal appear to be very different from the main Terminal 2 and Terminal 3. Although the focus of this study was on the main Terminal 2 and Terminal 3, a site visit of the Budget Terminal provided a better understanding of the airport operations from a different perspective. This has helped to trigger the identification of some JIT principles that appeared to have been incorporated into the operations of the main Terminal 2 and Terminal 3 but were not recognized as such earlier on.

During the third visit in August 2008, a briefing was also conducted for the research team that centered mainly on the airside area that included the Baggage Handling System (BHS) and airfields in both Terminal 2 and

Terminal 3. Other areas were also covered and explained to ensure a comprehensive understanding of the airport operations, especially in relation to areas that were missed out in the previous two visits.

## Interviews

Interviews are important to obtain first-hand anecdotal information from respondents with the necessary experience in the chosen field of study (Kumar, 2005). Apart from the site visits, interviews were also significant for this study in drawing out the experiential based opinions relating to the applicability of JIT principles in the airport.

### *Justification for interviews*

In the context of this study, the purpose of the interviews was to gain deeper insights to specific issues and areas which may have been overlooked during the observational walk-through. At the same time, interviews with personnel other than those who led the site visits would generate additional knowledge and valuable opinions. These inputs as well as the questions asked were significant in contributing towards quality and a more comprehensive set of findings.

### *Conducting the Interviews*

The interviews were conducted in July and August 2008 with two experienced managers from the relevant airport authorities. The first interviewee was a master airport planner who was familiar with the design and planning of airport terminals and the associated facilities, the design and planning of airfield facilities (e.g. runway, taxiway and apron) as well as the development of the airport master plans. The first interview focused mainly on the planning issues relating to terminal facilities such as the check-in area, immigration, security system, gate holding rooms, etc. In the interview, in-depth comparisons were frequently made between Terminal 2 and Terminal 3 to distill the differences between these two terminals.

The expertise of the second interviewee was in passenger terminal planning. Equipped with an architectural background, the second interview session focused on design aspects that were not covered previously. Valuable lessons were provided through the photographs shown by the interviewee on all three terminals which explained how the Changi International Airport has progressively evolved over the years since the commencement of Terminal 1.

## Questionnaire Survey

Quite apart from the site visits and interviews, the questionnaire survey offered yet another research tool to reach out to a larger number of people who can provide inputs to the subject matter of the study (Kumar, 2005).

### *Justification for Questionnaire Survey*

Site visits and interviews remained insufficient to find out the applicability of the JIT concept in the airport. More inputs were needed to reflect on how the rest of the airport employees feel about the improvements that can still be made to the future terminals or to maintain the excellent service quality achieved in Terminal 2 and Terminal 3. With this in mind, a questionnaire survey was conducted with 33 persons who were employed in the airport.

### *Profile of Respondents*

Two organizations who were directly responsible for managing and overseeing the airport participated in the surveys. The respondents were employees involved with airport planning and development, airport operations, airport management, airport commercial and retail operations, aviation security, engineering and real estate development as well as corporate strategy and business development. The respondents were mainly managers and senior managers in the airport. Most of them have had at least two years of experience in the aviation industry.

### *Conducting the Questionnaire Survey*

Understanding that a large number of the respondents may not have heard of the JIT concept, a brief introduction of each of the seven JIT principles was first made to the respondents, a sample of which is shown in the survey questionnaire attached in Appendix 1. In addition to understanding the JIT principles, the respondents also need to be familiar with the airport operations and the associated features. Because pictures speak a thousand words, power-point presentation slides were also used to illustrate each particular aspect of the airport that was highlighted together with the relevant JIT principle(s). These power-point presentation slides are attached in Appendix 2. Finally, the survey questionnaire was completed with an example that showed the thinking process of how the respondents should go about answering the questions.

For the questionnaire survey itself, the respondents were required to rate (on a scale of 1 to 5, with 1 being the least applicable and 5 being the most applicable) the applicability of each of the JIT principles to each of the airport elements. A choice was also given to the respondents to not rate at all if in their opinions, the JIT principles were not applicable in the particular case. Not all the JIT principles were evaluated in the questionnaire survey due to the large number (31) of airport features to be examined. In reality, the actual number of airport features will add up to 49 items. However, the 18 elements excluded from the questionnaire were relatively insignificant. The reason for exclusion was either the principle displayed in the airport element may be too evident or the principle was repeated in a similar fashion in another significant airport feature.

*Chapter 5*

# JIT PRINCIPLES FOR SPATIAL LAYOUT

## JIT PRINCIPLES IN THE AIRPORT CONTEXT

The application of JIT principles is not restricted only to the construction or manufacturing stage. It can also be extended to the planning stage in the layout of a building. However, certain JIT principles may need to be redefined for application, depending on the functional operations and needs of the business in the building. The seven JIT principles presented earlier will be examined in the context of the Changi International Airport.

### Principle 1. Elimination of Waste in the Airport

Elimination of waste refers to eliminating all unnecessary stocks and to decrease or minimize wastage (Low and Mok, 1999). In the context of the airport terminal, it can be referred to the minimizing of waste as applied to the airport layout and its work processes. The work processes are the repetitive and routine procedures required to perform passenger check-ins, to seek clearance through immigration, to board the planes and so on. The seven causes of waste under the elimination of waste principle are evaluated below. However it should be noted at this stage that not all of these may be relevant in the context of an airport.

#### *Waste from Over-Production*

Waste from over-production in manufacturing would typically refer to the excess quantities of the manufactured products. In this context, demand can be quantified quite easily. If the supply from the production exceeds the amount of demand, there is over-production. But this may not be identifiable so readily in the airport context as the object to be quantified is service. And as is commonly acknowledged, service can be a subjective component which has no absolute mode for calculation, where the limits are not possible to be defined, let alone to determine at which point over-production is deemed to occur. However, boundaries of comparisons can be assumed from the demands of the users during the peak and off-peak periods in the airport. For example, at times when there are low frequencies of flights, there would be fewer people using the lifts. When there are more flights, there may be many more people who may want to use the lifts at any one time. Sometimes, people may prefer to use the escalators instead of the lifts. It appears difficult to predict the actual or estimated amount of production that should be supplied. Consequently, the nature of this waste may seem difficult to encapsulate and appears therefore irrelevant in the context of the airport.

***Inventory Waste***

Similarly, inventory waste does not seem to be applicable in the context of the airport. This is because inventory storage is required to maintain quality service standards for passengers arising from the demand uncertainties in the airport.

***Waste from Product Defects***

Unlike manufacturing, there are predominately no physical products associated with an airport building *per se* where quality service provision is concerned. The provision of excellent service for the purpose of serving the passengers appears to be the key emphasis of an airport. Even if service can be termed as a product, service itself can be subjective and can differ from passenger to passenger. Hence, it appears difficult to judge off-hand if the service provided is considered "defective". Consequently, waste from product defects does not seem to be relevant where applicability in the airport context is concerned.

***Transportation Waste and Processing Waste***

The airport does not operate like a factory where components form part and parcel of the final products. Hence transportation waste and processing waste arising from the physical production process do not appear to be of concern to airport management in terms of the spatial layout design within the airport terminals.

### *Waste of Waiting Time*

Waste of waiting time refers to the overall waiting time experienced by the passengers. This is related to the number of mandatory work processes that each passenger has to go through. The waiting time can be quantified during peak and off-peak periods. There may be other factors that will affect the duration of the waiting time. Typically, a passenger is observed and/or expected to wait longer during the peak period. The concern should therefore be focused on the methods of how to minimize the waiting time so as to achieve efficiency that translates to good quality service.

### *Waste of Motion*

In achieving good spatial layout design, constant efforts are needed to plan the airport layout and/or architecture to accommodate the mandatory work processes in a more efficient manner. For example, airport gates are commonly situated away from the immigration gates so as to prevent congestion. There are also other creative ways where the airport has deployed in managing such waste. In many cases, the use of appropriate technology has been seen as a good tool for implementation to minimize and/or eliminate waste of motion and unnecessary efforts.

### *Waste of Space and Energy*

Waste of space and waste of energy are two relevant causes of waste in the context of the airport. Waste of space refers to any wastage of space that occurred as a result of poor design in the spatial layout. This can be avoided if the space is planned properly and flexibility is exercised for the space to be fully utilized.

Waste of energy refers to any form of energy waste. This predominately comes from Mechanical and Electrical (MandE) services that include lighting, air-conditioning, lifts, escalators and others. Given that the present emphasis in construction is on sustainability, this form of waste will be of importance to the management of Changi International Airport. Some possible ways to eliminate this waste can include using energy-saving devices or to position the MandE services near to the occupants. For example, placing the air-conditioning diffusers nearer to the occupants at ground level in an airport terminal would help to save more money than to supply cool air from a high ceiling level.

## Principle 2. Kanban System in the Airport

Some modifications need to be made in order to fit this principle appropriately within the airport context. Since the work flow in the airport is not about physical construction or production process, the workstations can be referred to as procedures that a passenger has to go through from one station to the next. Hence, the kanban or pull system can be defined to mean attracting or "pulling" a process or people over to the next station to achieve benefits that focus on either the needs of the airport or those of the customers'. One useful example is the directional signage that is positioned at strategic locations to guide the passengers. This is especially crucial when there are choices to various locations which a passenger can head to next in the airport terminal.

The kanban system can also be defined as a principle that provides for a service when there is a demand for it. The service does not need to be produced and stored on the spot. But it will be supplied when the need arises. Some airport planning uses the push system and produce more than what is needed in order to maintain quality standards. This is undesirable. Unlike manufacturing, the demand for certain things (especially services) cannot be predicted that readily. One such case relates to the trolleys used in the airport terminal. It appears difficult to determine whether a passenger will pick up a trolley to use as this action is dependent on the passenger's preference and the amount of baggage carried. To provide good quality service, excess trolleys have to be provided because the airport management cannot have less than what is necessary for the passengers.

## Principle 3. Uninterrupted Work Flow in the Airport

In the context of an airport, the focused factory concept can be related to the grouping of a single family of items and/or processes arranged together in the layout of an airport terminal. On the other hand, the group technology concept can be described as the grouping of family of similar parts or components in a continuous layout that leads from one passenger procedure to another. For example, the airport displays the group technology technique of providing a layout that leads passengers from the check-in area to the immigration area.

Much gain can be derived from the group technology technique. When implemented, the facilities layout should be able to facilitate ease of movement of work flow, people and information (Low and Mok, 1999).

Coordination and smooth work flows can be achieved. There is also minimal wastage of time when items or equipment are arranged in a continuous, smooth sequence. The result would lead to higher quality service (Low and Chan, 1997).

For design of simplification and automation, the airport terminal can make use of this technique to "install" the easiest possible route for any passenger to go through without difficulty and without the help of signage. In addition, automation had proven to be strategically useful in the Changi International Airport.

The technique of reduction of process set-up time does not appear to be that relevant in the airport. This is because airport processes are not concerned with the manufacturing of products which would involve the tooling of machines. As for the Total Preventive Maintenance program, maintenance issues are outsourced to the operators and suppliers. Penalties can be imposed if breakdowns happen to the People Mover System (PMS) and other equipment. This imposition is necessary because the service provided by these machinery and equipment is highly valued in the airport.

## Principle 4. Total Quality Control in the Airport

In the airport context, quality control means that the responsibility to ensure excellent work quality is delegated to all the employees. The employees play an important part in exploring ways, through their brainstorming sessions and discussions, to achieve excellent work standards through quality control circles. Quality in the airport means that quality is placed under control with careful planning and design of the spatial layout for each of the passenger procedures. This design can affect the work processes in terms of efficiency and the level of comfort for the passengers. When the airport management had taken steps to ensure quality in the work processes or spatial arrangement, quality control has to a large extent been exercised.

## Principle 5. Top Management Commitment and Employee Involvement in the Airport

To derive the most benefits from a good spatial layout in the airport, employees at all levels must be involved. Airport leaders and top management must believe in consulting widely and appreciate the need to work closely together with employees and customers to ensure the best possible operating structure for the airport (Ministry of Transport, 2008). Feedback and suggestions are to be obtained from both the employees and the airport users. The employees involved must cut across all hierarchical levels that range from those in airport management, operations and especially so for those who are at the frontline in the daily operations. Similarly, in certain areas pertaining to planning and development issues, employees should be encouraged to be actively involved and engaged.

In addition, in the case of the Changi International Airport, airport-wide campaigns and training programs have been conducted to train and motivate frontline staff to provide good customer service. The training is provided by the airport management to continuously develop the skills and flexibility of the airport employees.

## Principle 6. Supplier Relations in the Airport

The service excellence in an airport is dependent on the quality provided by the suppliers. Any external party who provides the airport with a service can be considered as a supplier. These suppliers include the maintenance contractor, People Mover System (PMS) provider, cab operators and so on. The JIT concept requires the suppliers to provide consistency and high quality in their performance or service. In addition, to maintain good airport quality, the suppliers should be producing quality products in the right quantities at the right time to suit the customer needs. If the suppliers do not perform up to the standards required, penalties can be imposed on them.

## Principle 7. Continuous Improvement in the Airport

This principle can be used to improve productivity through the layout of the terminal. In the context of the Changi International Airport, lessons learnt from Terminal 2 will be examined and used for improvements in Terminal 3 where appropriate. In addition, areas that have performed well should extend its applicability and continuity to subsequent terminals. One important reason why the Changi International Airport has been able to maintain its status as a world-class airport for many years is that steps have always been taken to strive for continuous improvements.

The second objective of examining the application of the JIT principles in each airport feature is reviewed below.

## RESULTS OF FINDINGS

Before the review, some assumptions underlying the evaluation need to be made. One key assumption is that all the activities and airport features mentioned in the findings are those which a typical economy-class traveler will need to pass through. Another important assumption is that the information obtained to document the findings was correct at the time of the study as this was based on sources such as the experiential observations of the research team, websites, books and evaluation from the three research methods (site visits, interviews and questionnaire survey) adopted for this study.

## DEPARTURE CURB SIDE

Figure 5.1. Curb side of Terminal 3.

The curb side (see Figure 5.1.) is defined as the area consisting of a sidewalk, covered or partially enclosed, bordering the terminal road system with an adjacent paved area that permits vehicles to off-load or load passengers (IATA, 2004). Several airport elements which have deployed the use of the JIT principles are discussed below.

## Signage

As departing passengers in the vehicles approach the Departure Hall of Terminal 3, they are able to see the signage of the entrance gate numbers and the respective airlines from afar. The kanban system is deployed here as the signage is provided just as it is needed for the passengers. This is facilitated by the layout of the signage and entrance doors which are specially designed to protrude and are positioned at an angle that are easily identified from afar. Upon reaching the curb side, passengers are better able to stop progressively at the respective passenger departure curb entrance door instead of coming to an abrupt halt suddenly. The signage is displayed prominently at the right time when passengers require directions as to where their vehicles should exactly stop. This leads to the next principle of continuous improvement discussed below.

The scenario described above for Terminal 3 appears to be an improvement which was made from Terminal 2 (see Figure 5.2). In Terminal 2, the signage is smaller and the overall image looks more cluttered when compared with the signage shown in Terminal 3. Furthermore, the entrance door in Terminal 3 was designed to show the gate door number clearly. Hence, at one glance, the gate numbers and airlines can be recognized easily. In addition, clocks are provided in Terminal 3 for passengers to keep track of time.

Figure 5.2. Curb side of Terminal 2.

## Roof

The roof is wide and long to deal with bad weather conditions such as from the extreme hot sun and heavy rain. This can be seen as a quality control measure because the roof will shelter all alighting passengers from bad weather conditions. The customer welfare is considered not only inside the terminal but also outside, from the moment when they first arrive at the airport.

Figure 5.3. Ramp outside main terminal.

## Ramp

The sidewalk is lined by ramps (see Figure 5.3) for passengers who are physically challenged and with limited mobility. At the same time, these ramps can also be used for passengers to roll their baggage up to the ground level from the road instead of having to lift their baggage one level up to the curb side. These ramps can only be found in Terminal 3 and appear to be an improvement from Terminal 2. The enhancement seems to come about because of continuous improvement.

## Trolley Carts

Trolley carts were outsourced to external suppliers. In this context, supplier relations are important to ensure that the supervision of good service quality be kept at an optimal standard. The supplier's responsibility is to ensure that sufficient trolley carts are always provided and are properly maintained. The JIT principle of uninterrupted work flow is applicable here for trolley carts to be provided near to all entrance doors where the passengers can have easy access to these carts upon reaching the airport. If all these trolley carts were to be placed together in a central location like the trolleys in supermarkets, then the passengers will need to walk to the central area to get to the carts. This will actually disrupt the pathway and result in interrupted work flow instead.

## Double-Façade Entrance Doors

For normal air-conditioned buildings, in many instances, there is only one automated layer of glass for the main entrance door. Once the main entrance door opens, it allows heat from the outdoor environment to mix with the air from the cool indoor environment. Even with the use of an air curtain, the internal cold air also leaks out externally. Passengers entering the airport terminal will therefore still feel the heat near the entrance area. There is a sudden change in temperature experienced by the users entering the terminal building from outside. This also results in high mechanical cooling requirements and thus, leads to high energy consumption and costs.

For the airport in Changi, cool air is contained within the indoor environment as there are two pairs of sliding doors (see Figure 5.4) which the users will need to go through before entering the main terminal building. This seems to mirror the application of the JIT principle of total quality control of the indoor air environment to create a comfortable and cool environment for the passengers. At the same time, this airport feature also helps to eliminate energy waste because the thermostat control can now be adjusted upwards than previously where a lower temperature needs to be set to cool the main entrance area.

Figure 5.4. Two automated doors at entrance to terminal building.

## Cable-Net Wall

Figure 5.5. Cable-net wall (glass façade at entrance).

The cable-net wall is the spectacular glass façade (see Figure 5.5) at the curb side. This improved glass façade helps to facilitate maximum light transmission into the interiors of the terminal building but yet reduces the direct heat effect of the sun. Energy waste is reduced because more daylight is allowed into the terminal building. Consequently, the use of artificial lighting can be reduced.

Eliminating energy waste also helps to reinforce the JIT principle of quality control. Apart from exercising quality control in saving energy, other benefits can also be achieved. The architectural objective emphasized the idea of twinning maximum transparency with awareness of spatial fluidity and visual continuity for the users to see ahead of places (Changi Airport, 2008). For example, the check-in counters inside the terminal building can be viewed from the external drop-off area. In addition, the lower three meters of the facade is specially made clear for the users to see through the external envelope wall.

Due to the extensive use of glass, glare may be an issue affecting the counter staffs using the computer monitors at the departure check-in area. To counteract the potential glare problem, fritted glass panels on the façade were used. Since this is a full height glass façade and glare will pose a stronger impact internally at higher heights, the fritting intensity is increased accordingly with the height of the glass panels to minimize glare. In addition to its main function of accomplishing maximum day-lighting, it has two additional functions described above which demonstrated the JIT principle of total quality control when the designers planned for the use of the glass façade. The same piece of façade can cater to different purposes to improve the visual experiences of the passengers, the ergonomic aspects of the employees (from glare) and the overall well-being of the users.

## Summary for Departure Curb Side of Terminal 3

The results from the questionnaire survey are summarized in Table 5.1. The applicability of each JIT principle is based on the questionnaire results which will be discussed in the evaluation and recommendations section in the next chapter. From the survey results, appropriate actions are suggested to assess whether the features can be further improved. This is especially important for improving those features with perceptibly low applicability and where recommendations can further be made for improvements. The details of these recommendations will be discussed in the next chapter.

**Table 5.1. JIT principles as applied in the departure curb side of Terminal 3**

| Airport features | Applicability of JIT principles | Applicability | | |
|---|---|---|---|---|
| | | High | Neutral | Low |
| Signage | Kanban/Pull System | | ✓ | |
| | Continuous improvement | ✓ | | |
| Roof | Total Quality Control | | ✓ | |
| Ramp | Continuous improvement* | - | - | - |
| Trolley carts | Uninterrupted work flow | ✓ | | |
| | Supplier relations | | ✓ | |
| Cable-net wall | Elimination of energy waste | | | ✓ |
| | Total Quality Control | | ✓ | |
| Double-façade entrance doors | Elimination of energy waste | ✓ | | |
| | Total Quality Control | | ✓ | |

(*Excluded in the questionnaire as this seems to be an evident application of the principle.)

## Departure Curb Side Atrium

Figure 5.6. Atrium in Terminal 3.

The curb side atrium (see Figure 5.6.) refers to the area where the passengers first step into the departure hall from the curb side. Upon arrival at the atrium, the passengers are welcomed by the dazzling environment and ambience delight of what Terminal 3 has to offer. One key contribution towards this delightful experiential effect comes from the roof.

Figure 5.7. Smart roof and its 919 skylights in Terminal 3.

## Smart Roof

The applicability of the JIT principle of total quality control appears to be demonstrated in many areas through the smart roof of the terminal building. The butterfly roof design incorporated several functions in which relevant JIT principles may be applied. Firstly, 919 skylights (see Figure 5.7) were installed throughout the entire roof. These skylights are equipped with specially designed reflector panels which automatically adjust to the prevailing sunlight and allow the right amount of natural and uniform daylight into the terminal building (Airport Technology, 2008)

Besides the reflector panels, the butterfly louvers (see Figure 5.8) above the roof can also be adjusted accordingly for good day-lighting and appropriate heat conditions. The louvers adjust themselves according to the sun's position and the current light intensity experienced (Lim, 2008). In less-sunny weather conditions, these mechanically operated louvers will flip open to receive the maximum amount of day-lighting. If the sensors detect hot and sunny weather conditions, the louvers will close to limit the amount of daylight and heat entering the building.

Figure 5.8. Butterfly louvers on top of Terminal 3.

At the same time, the butterfly louvers can also act as a secondary roof. As a secondary roof, it shades and protects the primary roof membrane against water tightness problems (Lim, 2008). Lastly, the amount of daylight is controlled by the louvers using its computerized system for various zones. One such zone is the greenery zone found in Terminal 3 where plants need natural daylight to sustain and grow.

In addition to the use of such technology to control the amount of daylight entering the terminal building, tube/interior lights are also installed to illuminate the departure hall during the night when there is no sunlight or to make up for insufficient daylight during cloudy days. Lest there is loss of power or malfunction, there is also a back-up system to ensure that the terminal building can continue to function efficiently. This helps to ensure total quality control of the lighting condition in the airport terminal building because the light intensity will be kept consistently the same regardless of the outside sun conditions. This also helps to provide a soothing ambience at all times of the day and night in the building (Changi Airport, 2008).

Besides exercising the JIT principle of total quality control, the elimination of energy waste principle is also applied for the cable-net wall where the double-insulating glass protects the skylights against the weather. Low-emissivity coating is applied to the glass to admit 70% of visible daylight and only 35% of thermal heat. Admitting more daylight but less heat helps in energy saving. In addition, this not only allows natural daylight into the building, the lighting is also directed all the way to the basement areas such as the baggage claim area. With the use of natural daylight, energy consumption for artificial lighting can be lowered. Hence, this helps to eliminate the wastage of energy and reduce the operating costs (life cycle costs) for the airport.

## Transparent Pavilion

The transparent pavilion refers to the large pavilion-like design for the terminal with transparent internal facades. This feature provides the users with a generous feel and sense of space (Lim, 2008). The JIT principle of kanban/pull system appears to apply in this case because the users will subconsciously know in advance which direction to proceed to next due to the extensive use of transparent glass inside the building. The facilities ahead of the passengers are actually pulling them to walk towards these facilities because everything can be openly seen beforehand.

The spatial layout of Terminal 3 is such that it allows a continuation of processes that progressively leads the passengers from one area to another area. This entire process consists of the progress of the passengers from the drop-off points to the boarding gates. This seems to mirror the JIT technique of group technology under the JIT principle of uninterrupted work flow. The open and transparent view allows the passengers to navigate smoothly inside the terminal building without disruption or the need to stop to ask for directions.

## Lifts, Escalators and Travelators

Lifts, escalators and travelators (see Figure 5.9) are useful for departing passengers who may have ventured to the wrong level of the building when they first stepped into the airport terminal. The passengers may be at the arrival hall level when they need to be at the departure hall level. Hence, these transport equipment must be located at the atrium so that the passengers can proceed to the correct level quickly and conveniently. Passengers are inevitably concerned with boarding their planes on time. This appears to mirror the JIT principle of eliminating time waste for such passengers whereby the transport equipment which is appropriately positioned can help to move them along smoothly and to reduce their anxiety.

The principle of eliminating energy waste is also mirrored. There are motion sensors installed for the escalators and travelators for these to be activated only when movements are detected. At other times, the equipment will stop operating to conserve energy. Without these motion sensors and under normal circumstances, the equipment will continue to operate regardless of whether there are people using them. Hence, the use of energy-saving motion sensors will help to eliminate energy wastage.

(a) Travelators

(b) Lifts

(c) Escalators

Figure 5.9. (a), (b) and (c) Vertical transportation in Terminal 3.

In the Changi International Airport, the layout is designed to adopt a single storey workflow for the passengers throughout the entire immigration-to-boarding process. This means that as far as possible, departing and arriving passengers need not change level vertically in their respective passenger flow routes. This is especially so since both the departure hall and the arrival hall occupies a single storey each. Disruptive level changes can create annoyance and confusion to the passengers. This spatial layout is also more user-friendly for the physically challenged or passengers with mobility problems. It would appear that the JIT principle of quality control has been exercised in this area. It is also noteworthy that in the event of breakdowns, alternative transportation can be used by the passengers.

### *Signage*

The functional principle for signage is directional clarity. The fundamental concept of signage is about letting the passengers know where they are relative to the entire terminal building and to clearly see where they are heading or intending to go to next (IATA, 2004). With good signage, the passengers will be more aware when they move from one processing zone to another if they are not familiar with the place or are unable to visualize clearly from the building layout where they are and where they need to go to next. In fact, the terminal building itself should indicate clearly the path of travel (IATA, 2004). While a good airport layout design is one that can rely minimally on signage, nevertheless, there are still instances where the signage is needed. Two types of signage in Terminal 3 will be discussed.

The first type of signage is the normal signage (see Figure 5.10) which can be seen throughout the entire terminal and which adopted a color scheme to distinguish between the various facilities. For example, the signage in blue is related to passenger facilities. Subsequently, this signage can be found throughout the passenger flow process route at critical points where directional changes and decisions have to be made. Passengers will then know exactly where to proceed next just in time before the next procedure surfaces and immediately after the preceding one has been completed. The kanban system in the JIT concept is mirrored here since this signage will always be provided at a timely manner when needed or when in demand at the appropriate times.

Figure 5.10. Signage near check-in hall of Terminal 3.

## Summary for Departure Curb Side Atrium

The findings from the questionnaire survey relating to the applicability of the JIT principles for the departure curb side atrium is summarized in Table 5.2.

**Table 5.2. JIT principles as applied in departure curb side atrium**

| Airport features | Applicability of JIT principles | Applicability | | |
|---|---|---|---|---|
| | | High | Neutral | Low |
| Butterfly roof | Elimination of energy waste | ✓ | | |
| | Total Quality Control | ✓ | | |
| Transparent pavilion | Kanban/Pull System | | | ✓ |
| | Uninterrupted work flow – group technology | | | ✓ |
| Lifts, escalators and travelators | Elimination of time and energy waste | | ✓ | |
| | Total Quality Control | ✓ | | |
| Signage | Kanban/Pull System* | - | - | - |

(*Excluded in the questionnaire as this seems to be an evident application of the principle.)

## CHECK-IN HALL

The acceptance by the airlines of passengers and their checked baggage take place in the check-in hall which consists of the check-in counters with appropriate baggage conveyance facilities. The layout of the check-in hall is largely influenced by the check-in concept planned by the airport authorities. It is therefore essential that the airlines or their handling agents are consulted early in the terminal design and planning process. This will ensure optimal operational effectiveness in the allocation of check-in counters to the various airlines and their handling agents (IATA, 2004). The air-conditioning system, check-in counters, baggage system and other features in the check-in hall will be discussed relative to the applicability of the appropriate JIT principles.

### Pinnacles – Signage with Air-Conditioning Diffusers

The air-conditioning units are well hidden at the check-in hall in the terminal building. These units have been intelligently integrated with the check-in counters. Each of the counter signage is known as a pinnacle (see Figure 5.11). The pinnacle is the structure which displays the luminous check-in gate number that the passengers look out for when they check-in. The structure has dual functions.

Figure 5.11. A typical pinnacle in Terminal 3.

Firstly, the pinnacle informs the passengers from afar where they are supposed to go for their check-ins. The tall and luminous structure gives a distinctive view of the gate number (example, No.11 in Figure 5.11) at one glance. There is space saving because the air-conditioning system is also built into this architectural feature.

Secondly, each pinnacle incorporates the air-conditioning diffusers. These air conditioning units are also strategically placed throughout the terminal building to provide a consistently cool temperature at the occupancy level. Cold air which is supplied at a high ceiling level will waste energy in cooling large areas of unoccupied space. This new displacement air-conditioning system is used in the departure hall of Terminal 3 to cool the occupied area at about five meters above the floor level where most of the occupants can be found. Hence, the pinnacles help to eliminate energy waste which is reflective of the JIT concept.

At the same time, from the aesthetical point of view, having the conventional air-conditioning system with ducts and vents in the ceiling would ruin the architectural beauty of the roof (Lim, 2008). These design principles appear to have complemented each other very well and added tremendous benefits to the Changi International Airport.

## Flight Information Display System (FIDS)

Situated at the entrance of the check-in hall are the Flight Information Display System (FIDS). Previously, the FIDS in Terminal 2 consisted of a big board (see Figure 5.12) located in a centralized position with six door gates sharing two such systems. This often caused human congestion at these areas. Continuous improvements appear to have been implemented since then in Terminal 3 with the FIDS screens (see Figure 5.13) situated along the passenger route to the check-in hall with two door gates sharing one system. Congestion is now no longer an issue. Through such continuous improvements, uninterrupted work flow has also been achieved since passengers no longer have to walk to the centralized FIDS board before proceeding to the check-in counters. There is overlapping of two JIT principles relating to continuous improvement and uninterrupted work flow, resulting in better outcome for the airport users.

## Check-in Counters and Layout

Terminal 3 uses the island type of check-in layout. This type of layout means each island, where the axis is oriented parallel to the flow of passengers through the terminal concourse, consists of about 10 to 20 individual check-in counters on each side (IATA, 2004).

Figure 5.12. FIDS in Terminal 2.

Figure 5.13. FIDS in Terminal 3.

This facilitates smooth passenger flow. The location of the airlines in this layout indicates that the focused factory technique under the JIT principle of uninterrupted work flow is deployed because the airline counters are positioned together at the check-in hall.

There is also sharing of check-in counters by airlines under the same alliance umbrella. In this instance, the passengers checking into Singapore Airlines flights may also use other allied airline counters. In addition, in the check-in hall, lighting, air-conditioning outlets and baggage conveyor tracks are centralized and contained within the check-in islands. These are not scattered around. This one-location based streamlining of services required a high level of strategizing at the design stage and appears to mirror the focused factory technique.

Design for simplification and automation was also reflected in the check-in area. Simplification comes when there is more than one way to perform the check-ins. Passengers can choose between manual check-in, internet check-in, phone check-in and self check-in. With advanced technology through the facilities installed, passengers can do their check-in over the telephone, short messaging service (SMS), facsimile check-in service and check-in via the internet (Changi Airport, 2006). The passengers can then just simply collect their boarding passes at the respective airline counters and at the time specified by the airlines. Procedural matters have now been simplified and left to automation to help the passengers complete the check-in procedures smoothly.

The JIT principle of quality control is also reflected in the check-in area. According to the IATA airport development reference manual, the operations in Terminal 3 strived to achieve a good level of service and comfort for the passengers by allocating a comfortable amount of space per passenger with first-rate service quality (IATA, 2004). For example, in meeting the requirements for a world-class passenger terminal like Terminal 3, the amount of space allocated to a standard passenger with a trolley cart should be in the range of between $1.3m^2$ to $1.8m^2$. There should also be space allocated for electronic self check-in kiosks, including space for baggage acceptance. Apart from the sufficient space allocated for the passengers, there should also be large circulation areas and adequate seating to cater to the senders while waiting for the passengers near the check-in facilities. All these have been provided for in Terminal 3.

## Baggage Handling System (BHS)

From the walk-through and interviews which formed part of the empirical portion of this study, it was noted that there are only two sections of BHS in Terminal 2, which are namely the weighing and tagging section and the induction section respectively. Weighing the baggage will help check for the overweight and odd-sized baggage. Tagging the baggage means attaching the baggage with an identification label. The label allows airport employees to identify the baggage origin of departure, destination and flight code which will be translated into a bar code for the BHS. This will facilitate the movement of baggage to the respective aircrafts and for identification purposes. The label is especially useful in locating lost baggage almost instantly. Induction means sending the baggage for security checks and sorting before transferring to the respective aircrafts eventually.

In Terminal 2, the processes of weighing and tagging are undertaken together in the same section. When there are numerous pieces of baggage to be weighed and tagged, the system cannot proceed on to the induction stage because only one baggage is allowed to be inducted at any one time. To counter this, the check-in desk employees have to move the baggage out of the section so as to allow one baggage to be inducted at any one time. This appears to result in inefficiency and interrupt with the baggage flow process.

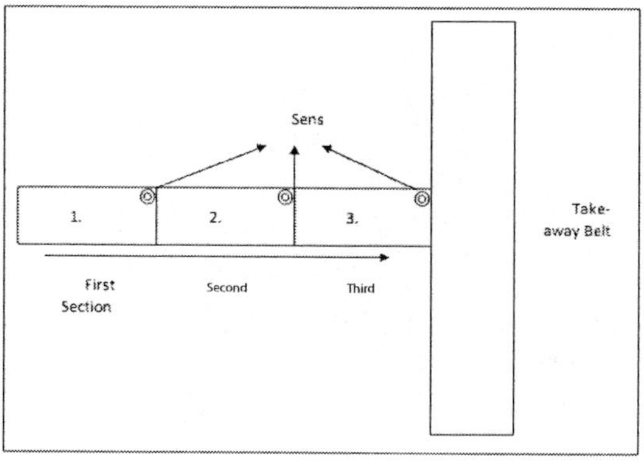


Figure 5.14. Check-in BHS in Terminal 3.

Incorporating the practice of continuous improvement, the BHS in Terminal 3 is divided into three sections as illustrated in Figure 5.14. The first section is weighing, followed by tagging and lastly, induction. The first section of weighing is to determine if the checked-in baggage is of an acceptable weight and size. If so, the sensor by the side will "approve" the process and the baggage will then proceed to the second section for tagging. In this section, the baggage will be tagged by employees before proceeding on to the induction stage.

For passengers who check in several bags, the tagging section can hold several pieces of baggage at any one time and the next ready tagged baggage can continue to go onto the induction section to be inducted into the take-away belt. Airline employees now need not get up from their seats to move the baggage unlike in the previous practice. Hence, in this case, there is elimination of motion (effort) waste by the employees.

After clearing the three sections of weighing, tagging and induction, the baggage will go on to the take-away belt to pass through the screening process. Thereafter, the cleared baggage will then move into the tilt-tray system to be transferred to the aircrafts. It can be observed that with this improved system, the processes in the BHS can be carried out continuously and without disruption.

In the BHS, it is noteworthy that a section of the conveyor belt system is slanted at an angle (see Figure 5.15). This is to facilitate smooth baggage movement by allowing the baggage to fall flat onto the conveyor belt. This was observed in the BHS in Terminal 3.

Figure 5.15. Conveyor belt system (three sections) in Terminal 3.

In contrast, airport employees in Terminal 2 have to get up from their seats to lay the baggage flat on the conveyor belt because the baggage might still be "standing" up straight after the induction stage. This is another example that reflects the JIT principle of uninterrupted work process for the airport employees. The improvement means that the employees need not waste unnecessary effort and motion. The enhanced feature does not interrupt the work of the employees by requiring them to constantly shift the baggage to lay the luggage flat on its side.

Another application of the JIT principle of continuous improvement is reflected in the odd sized belt. In Terminal 2, every four check-in counters have one odd sized belt. The odd sized belt refers to the belt which takes in the odd size baggage of an unusual shape, size or weight to the basement for sorting before being taken to the respective aircrafts for loading. However, in Terminal 3, every two counters now share the use of one odd sized belt which helps to enhance the efficiency of baggage handling.

Lastly, for passengers who wish to check-in their luggage early or have long connection times, an automated early baggage storage system is in place to store the luggage. Passengers are now able to check-in early, up to 48 hours beforehand. This reflects another example of the continuous improvement made. As similarly done in other previous sections, a summary table is shown in Table 5.3.

## SUMMARY FOR CHECK-IN HALL

The findings from the questionnaire survey relating to the applicability of the JIT principles for the check-in hall is summarized in Table 5.3.

**Table 5.3. JIT principles applied in the check-in hall**

| Airport features | Applicability of JIT principles | Applicability | | |
|---|---|---|---|---|
| | | High | Neutral | Low |
| Pinnacles | Elimination of space and energy waste | ✓ | | |
| Flight Display Information Board (FIDS) | Uninterrupted work flow* | - | - | - |
| | Continuous improvement | | | ✓ |
| Check-in layout | Uninterrupted work flow – Focused factory and Design for simplification and automation | ✓ | | |
| | Quality Control | ✓ | | |
| Conveyor Belt System | Elimination of waste of motion | | ✓ | |
| | Uninterrupted work flow* | - | - | - |
| | Continuous improvement | ✓ | | |

(*Excluded in the questionnaire as this seems to be an evident application of the principle.)

## RETAIL, RESTAURANTS AND OTHER CONCESSION AREAS

As a good practice, the airport employees at the check-in counters would have informed the passengers of how much time they have left or at what time they should clear immigration in time for boarding the aircraft. For example, if the flight is a highly secured flight, the passengers have to clear the immigration stage early because it will take a longer time than usual for the screening process to be completed. Hence, if the passengers find that they still have some time before clearing the immigration or wish to spend a longer time with their family members and friends, they can choose to stay in the concession areas at the landside of the terminal building.

## Shops At the Landside and Airside

The JIT principle of total quality control appears to have been exercised for the design of the spatial layout in the concession areas at the landside of the terminal building. The landside refers to the area in the terminal building that the passenger has access to before going through the passport control station. The airside therefore refers to the areas beyond the passport control station. In most major international airports, amenities and services have been strategically placed throughout the terminal building to give easy and convenient access to the products and services that the passengers may need. The facilities provided can include money changing, internet kiosks, massage chairs and even a fully stocked supermarket that are all arranged either together at the same level or in the same area. The passengers get to enjoy more than 100 retail and over 40 food and beverage outlets in the Changi International Airport with wide ranging services from both established local and international brands.

Retail shops can be found in both the landside and in the airside of the airport. A design principle that has been applied seems apparent in the planning arrangement. In this context, it is noteworthy that there are many more retail shops at the airside (transit mall) than at the landside (concession areas) of the terminal building. In fact, this appears to be a deliberate planning decision made by the airport planners.

In addition to the attractive revenue income, this design principle is also for crowd control purposes. As the airport planners do not wish to see a large surge of passengers going through the departure immigration gate at any one time, especially in the window periods which are close to the gate closing times, more shops were planned to be located in the transit mall on the airside. This will help to "pull" passengers who have arrived early to get their immigration procedures completed as early as possible.

By "pulling" these passengers to the airside, congestion at the immigration gate can be eliminated. The outcome is that the anxiety level of the passengers can then be better managed at a more comfortable and relaxed manner. The JIT principle relating to the kanban/pull system was reflected in this instance. Furthermore, to maintain and balance the total numbers of passengers in the landside and the airside, retails shops have also been placed at the landside to attract passengers to shop and linger in this area.

### Viewing Gallery

The JIT principle of quality control appears to be reflected in the viewing gallery of the airport. The northern, southern and western facades are enveloped with a conventional unitized aluminum profile glazing system that uses double-glazed glass panels (Lim, 2008). These panels were selected considering the optimal transparency and energy performance. One such glass panel comprises of a framed laminated glass with an inner layer of heat-toughened glass and an outer layer of toughened glass. This laminated glass was also chosen for its acoustical performance in shutting out extremely loud aircraft noise.

## SUMMARY FOR RETAIL, RESTAURANTS AND OTHER CONCESSION AREAS

The findings from the questionnaire survey relating to the applicability of the JIT principles for the retail, restaurants and other concession areas is summarized in Table 5.4.

**Table 5.4. JIT principles applied in retail, restaurants and other concession areas**

| Airport features | Applicability of JIT principles | Applicability | | |
|---|---|---|---|---|
| | | High | Neutral | Low |
| Retail shops and amenities | Kanban/Pull System | | ✓ | |
| | Total Quality Control | ✓ | | |
| Viewing gallery | Quality Control* | - | - | - |

(*Excluded in the questionnaire as this seems to be an evident application of the principle.)

# Departure Immigration and Passport Control Area

Good signage and lighting conditions are significant considerations for both the airport employees and passengers in the departure immigration and passport control area.

## Signage

Once they have cleared the immigration procedures, the passengers have a directional decision to make – either to turn left or right (as shown in Figure 5.16). The positioning of the signage comes at the right time and in the right place. The JIT principle relating to the kanban/pull system is also reflected in this case. There is large and clear signage at one glance to show the direction for the passengers to proceed to the correct gate lounge, showing the gate lounges A on the left side and lounges B on the right side from the direction of the passenger pathway after clearing immigration. The numbering of the signage also starts from 1 all the way until the last number to make it easy for the passengers to find their way.

Figure 5.16. Dimmed lighting used and signage.

### Lighting Conditions

There is also quality control for the appropriate amount of lighting in the different zones in the terminal hall. There appears to be different lighting requirements in each zone. To create a further relaxed ambience and a wonderful shopping experience for the passengers, the lighting used is dimmer (as shown in Figure 5.16) after the immigration stations than those used throughout the departure hall where the lighting needs to be brighter to better facilitate the check-in procedures.

Dimmer lighting can help the passengers to feel more relaxed and reduce their anxiety prior to clearing immigration requirements. In the check-in area, there needs to be sufficient lighting for the passengers to identify the relevant check-in counters. There is also glare-free lighting for computer monitors as well as the flight information on the plasma screens. Soothing light is provided for the baggage claim area.

## SUMMARY FOR DEPARTURE IMMIGRATION AND PASSPORT CONTROL AREA

The findings from the questionnaire survey relating to the applicability of the JIT principles for the departure immigration and passport control area is summarized in Table 5.5.

**Table 5.5. JIT principles applied in the departure immigration and passport control area**

| Airport features | Applicability of JIT principles | Applicability | | |
|---|---|---|---|---|
| | | High | Neutral | Low |
| Signage | Kanban/Pull System* | - | - | - |
| Lighting conditions | Quality Control | | ✓ | |

(*Excluded in the questionnaire as this seems to be an evident application of the principle.)

## TRANSIT MALL

The passenger flow paths, shops in the airside as well as relax and rest areas are significant considerations for both the airport employees and passengers in the transit mall area.

## Passenger Flow Paths

Following a preplanned arrangement, about half the passengers will be directed to either the left or the right area to avoid overcrowding on one side of the transit area. Hence, this means that the passengers will pass by at least 50 percent of the retail shops. Such a concept was planned carefully by the interior planner for the terminal by working closely with the airport commercial department and other retail consultants to optimize the usage of all the available facilities. The planner also availed the passengers of the opportunity to experience the interior design ambience for the transit mall in Terminal 3. This suggests that the JIT principle of supplier relations is important to facilitate close working relationships among the stakeholders to optimize the layout plan of the transit area. Bridging the two JIT principles, suppliers are important to bring about greater improvements in the terminal building.

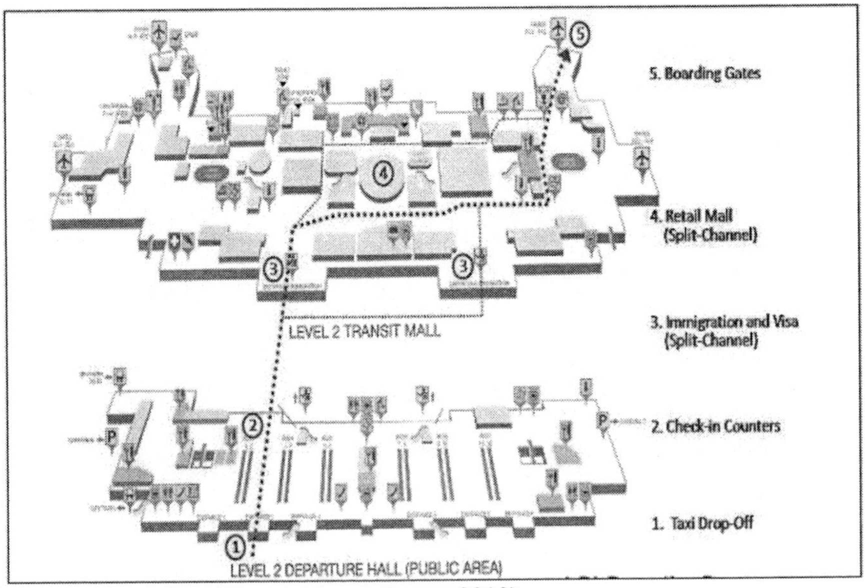


(Source: adapted from Changi Airport Group, 2010)

Figure 5.17. Passenger flow path in transit mall of Terminal 2.

Continuous efforts have been made to improve the passenger flow routes. As shown in Figure 5.17, from path route 3, there are split channels via path 3 that lead to path 4. This means that there are two paths that a passenger can take to walk through the transit mall. Passengers taking either path may possibly miss out some parts of the shopping experience. There is a hierarchy of pathways; namely the main primary pathway and the secondary pathway. Although path 3 is designated to be the primary pathway for passengers to walk through, some passengers may take path 4 (the secondary route) if they have been unclear of the directional flow and thus miss out on some of the shops along path 3. Beyond this, most passengers will simply walk to their boarding gates.

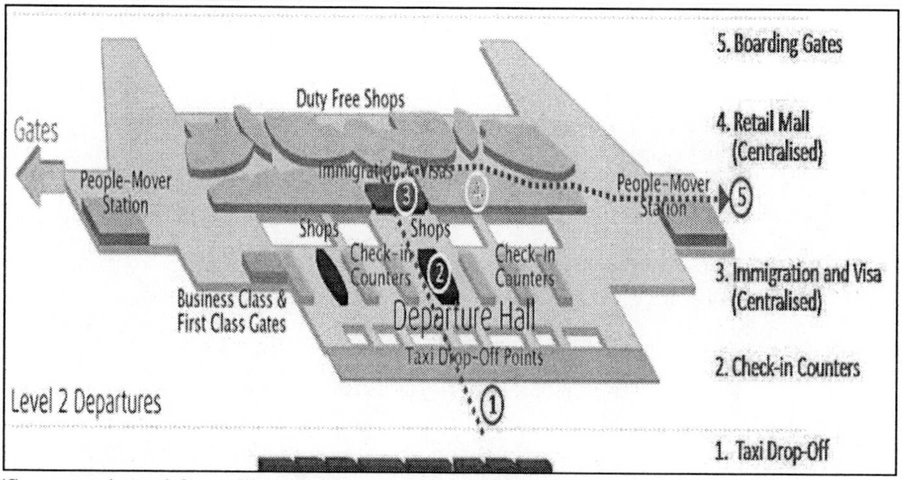


(Source: adapted from Changi Airport Group, 2010)

Figure 5.18. Passenger flow path in transit mall of Terminal 3.

The passenger pathway appears to be different in Terminal 3 as shown in Figure 5.18. In this case, there is only one obvious pathway, either on the left or on the right for the passengers to take. The JIT principle of continuous improvement was reflected in the comparison made above.

The passengers have two levels to choose after clearing immigration. The upper level houses most of the food and beverage (FandB) outlets and airline lounges. The lower level houses the retail outlets, entertainment activities and some FandB outlets. Because the passengers may proceed directly to the gate lounges, an attractive layout is therefore needed to direct them to the upper level. The JIT principle relating to the kanban/pull system is mirrored here.

There are visual aids such as dramatic lightings, advertisement billboards and graphics to catch the attention of the passengers. These make the passengers aware of the shops upstairs and provide clear visual cue to direct the passengers to the upper floors. As these visual aids are interesting, the passengers are attracted or pulled to the upper level.

Good illumination throughout the whole terminal (from the curb side to the aircraft) is important to the passengers for directional guidance. Instinctively, the passengers are directed to each station such as the check-in counters, the immigration counters, etc. The JIT principle relating to the kanban system is reflected through the use of suitable illumination levels that help to orientate as well as guide the passengers to their destinations.

## Shops in Airside

Figure 5.19. Design of zig-zag shop frontage.

To achieve the intended concept of finding one's way around easily as well as for clarity, the planner for the airport interiors also designed the layout of the shop frontage in a zig-zag manner (see Figure 5.19). This allows the passengers to see ahead to the next few shops from where they are. This may not be possible if a straight-line configuration has been adopted for the shops.

The zig-zag design will help the passengers in making their purchasing decisions and in determining whether they can get similar or other items in the shops ahead. The JIT principle of quality control is reflected here.

Furthermore, showing what are inside the shops from afar will help to enhance the attractiveness of the shops. This appears to be better than a layout design which will reveal the wares inside the shops only when the passengers are very near to the shops. In this context, the tenants in the shops in Terminal 3 can be given the chance to personalize their shops through their displays. They can be creative to make their shops stand out better than the other shops.

The group technology technique is also reflected here. There are more than 100 retail shops and more than 40 FandB outlets in the area. All the similar facilities are grouped together, complete with other entertainment facilities to create an exciting shopping experience for the passengers. All facilities are co-located, meaning these are eclectically mixed together and yet are able to create an attractive ambience. This helps to attract the passengers to walk through the mix of shops instead of passing them by. The variety of activities offered attracts the passengers to linger and to shop.

## Relax and Rest (Rand R) Areas

It is noteworthy that not all passengers would want to pass through the bustling activities provided by the shops. Some passengers may just want to rest before boarding their flights. As observed earlier, the lighting inside the transit mall has been dimmed. The lighting level in the Rand R areas is even dimmer to give a more relaxing environment for the passengers to rest. Similarly, the lighting level in the retail outlets is brighter to create an exciting and fun environment.

The outdoor terrace and pond (known as the Butterfly Garden) is one place where the passengers can go to for relaxation. The landscape architecture of Terminal 3 was designed to include a lush outdoor terrace and pond. This tropical retreat houses local plants and trees and is a welcome park-like setting for the passengers waiting for their flights (Changi Airport, 2008). In addition, there are a wide variety of facilities for the passengers to enjoy. These include free internet access, rest areas with snooze chairs, movie theatre, shower facilities, fitness and spa services. The JIT principle of quality control is well reflected here.

## SUMMARY FOR TRANSIT MALL

The findings from the questionnaire survey relating to the applicability of the JIT principles for the transit mall is summarized in Table 5.6.

**Table 5.6. JIT principles applied in transit mall**

| Airport features | Applicability of JIT principles | Applicability | | |
|---|---|---|---|---|
| | | High | Neutral | Low |
| Passenger flow paths | Supplier relations* | - | - | - |
| | Continuous improvement | ✓ | | |
| | Kanban/Pull System* | - | - | - |
| Design of shop frontage in a zig-zag manner | Total Quality Control | | ✓ | |
| | Uninterrupted work flow - group technology* | - | - | - |
| Rest areas | Quality Control* | - | - | - |

(*Excluded in the questionnaire as this seems to be an evident application of the principle.)

## GATE LOUNGES

Considerations for the gate lounges include the pathway to the gate lounges, decentralized screening and the gate holding rooms.

### Pathway to Respective Gates

The full shopping experience may be enjoyed to a greater extent by the transfer passengers who may have two to six hours to linger around in the transit area before proceeding to catch their next flights. The non-transit normal passengers usually have about 40 minutes to an hour before they board the aircrafts as most tend to arrive in time for check-in and other immigration procedures. Hence, it is quite likely that the non-transit normal passengers will proceed straight to the gate lounges and miss out on the shopping experience designed for them, especially if they are worried about being late for their flights.

In this instance, improvements have been made to the signage to provide more information to the passengers. The signage provides the maximum time needed (see Figure 5.20) for the passengers to walk to their respective gates. This additional piece of information allows the passengers to have a better control of their time. In addition, there are also fewer distractions around the signage which therefore allows the passengers to view the signage distinctively without wasting time. This practice mirrors the JIT principle of continuous improvement. Previously, smaller-sized signage and other advertisements or signage were placed together which can create confusion to the passengers when reading them.

Figure 5.20. Estimated time shown on signage.

Another improvement observed was made for the convenience of the passengers. For gates that are further away or at extreme ends of the terminal building, travelators and the People Mover System (PMS) are installed for the passengers to reach their gates quicker and more readily. The PMS will be further elaborated in details later.

## Decentralized Security Checks

Decentralized security screening is conducted at the gate lounges. Before the passengers proceed with their hand carry luggage to the boarding gate, security checks will be made for the hand carry luggage at the assigned boarding gates. The airport has consolidated the gate holding rooms in either pairs or groups (see Figure 5.21). The reason is that common gate holding rooms would require less manpower and security screening equipment. This arrangement can help to save capital costs as well as maintenance costs (Lim, 2008). Instead of having individual gate lounges, several lounges can be combined into one. There is elimination of space in line with the JIT concept. Less security screening equipment is needed as well.

Figure 5.21. Common gate holding rooms.

## Gate Holding Rooms

Figure 5.22. Gate holding rooms in "scallop" form in Terminal 3.

Figure 5.23. Straight gate holding rooms in Terminal 2.

To enable the passengers to locate their gate holding rooms readily, the "scallop form" (see Figure 5.22) of design was used for the glass walls to facilitate easy recognition of the gate holding rooms from afar. This design also makes walking to the gates less daunting because it creates an impression of a seemingly shorter walk to the gates. This appears to be different for Terminal 2 (see Figure 5.23) where the glass walls are lined straightly and the passengers may need to crank their heads to locate the individual gate numbers from a distance away. Hence, the JIT principle of continuous improvement appears to be mirrored in this comparison.

The quality of the gate holding rooms is important to the airport planners because the passengers may have to wait for about half an hour before they board the aircraft. Total quality control is to be exercised here. The Level of Service (LOS) standards set by IATA (2004) are complied to achieve an "A" grade. The standards take into consideration the sufficiency of seats, area of space provided per passenger, the facilities provided and so on. Sufficient seats are provided for the passengers according to the size of the aircraft. The larger the capacity of the aircraft, the more seats should be provided for the passengers in the gate holding room.

The basic quality provided for a gate holding room should be one which will not exacerbate anxiety and stress for the passengers. The interior design of the area is also taken into consideration to provide a stress-free environment for the passengers. Facilities like internet kiosks, televisions and telephones are provided as well. Artistic materials and designs are carefully chosen. An example of the seating arrangement in a gate holding room in Terminal 3 is shown in Figure 5.24.

Figure 5.24. Gate holding room and seating arrangement in Terminal 3.

Figure 5.25. Gate holding room in Terminal 2.

Seating arrangements in the gate holding room in Terminal 3 allow for the smaller and larger groups of passengers to comfortably stay together while waiting for their flights. This can be seen as a continuous improvement from Terminal 2 (see Figure 5.25). This principle of continuous improvement bridges with quality control to derive better value for the passengers. To help the passengers navigate to the correct gate, interior elements such as carpet patterns, seat colors and layout have been introduced to subtly demarcate the respective seating zones and the passenger flow corridors to the boarding gates (Lim, 2008). This aspect does not seem to come through that strongly in the design layout for Terminal 2. Beyond this point, once the boarding gates are open, the passengers are ready to board the aircraft to depart from the airport.

## Summary for Gate Lounges

The findings from the questionnaire survey relating to the applicability of the JIT principles for the gate lounges is summarized in Table 5.7.

**Table 5.7. JIT principles applied in the gate lounges**

| Airport features | Applicability of JIT principles | Applicability | | |
|---|---|---|---|---|
| | | High | Neutral | Low |
| Pathway to gate lounges | Continuous improvement* | - | - | - |
| Decentralized security screening | Elimination of space waste | | | ✓ |
| Gate holding rooms | Total Quality Control* | - | - | - |
| | Continuous improvement* | - | - | - |

(*Excluded in the questionnaire as this seems to be an evident application of the principle.)

## Arrival Area

As the arriving passengers first step out of the aircraft, they should find that the lighting is not too glaring for their eyes. The airport planners have implemented quality control for the lighting conditions in the terminal building. The dim lighting caters specially to this aspect of visual comfort for the passengers on their way out of the aircraft. Thereafter, the lighting gets brighter as the passengers move progressively to the immigration area, then to the baggage claim area and out of the airport. This is to let the passengers adjust gradually to the outdoor environment instead of exposing them to strong glare and brightness suddenly.

## Automated People Mover System (PMS)

The automated sky-train service can transport passengers between the terminals. There are three stations in Terminal 3 and two stations each in Terminal 1 and Terminal 2. The new stations and trains have Liquid Crystal Display (LCD) screens that inform the passengers where and when flights are arriving and departing as well as other important airport information. There is also ample space to accommodate the passengers and their luggage. The train cabins also have more vertical stanchions and handholds as well as designated areas for baggage trolleys to enhance the convenience of the users (Changi Airport, 2008).

Figure 5.26. Sky-train station in Terminal 3 (view from the public side).

Transfer passengers may need to make flight transfers from another terminal. Hence, they can make use of the sky-train service for this purpose. For security and commercial reasons, the passengers with boarding passes need to be separated from the other users of the airport for inter-terminal transfers. There are two main carriages. As shown in Figure 5.26, one side of the carriage will open for the passengers inside the transit area while the other side of another carriage will open to the public domain.

By using this system, there is no need to build two different sky-trains for the public users and the passengers that will account for high energy consumption and costs incurred. Having two separate sky-trains will also occupy a larger area in the need to have more tracks. If the users have to wait for only one sky-train that can be used by either the public users or the passengers at any one time, then the users on the other side will have to wait for another sky-train. This will increase waiting time. More importantly, this arrangement will ensure that there would not be a mix of passengers and public users for security purposes. As a result, elimination of waiting time, energy wastage and space wastage can be achieved.

## Immigration Area

Figure 5.27. Floor above the immigration area.

Figure 5.28. Elevated views of queues at immigration area.

As the passengers walk to the immigration area, they have to take an escalator down one level (see Figure 5.27). The JIT principle of uninterrupted work flow appears to be mirrored here in this layout. This is to facilitate the passengers to visually locate the shortest queue from a higher level when they travel on the escalator (see Figure 5.28). If the immigration area is placed at the same level, the passengers may find it difficult to distinguish which is the shortest queue at the eye level. Most likely, they may have to stop to decide which queue to join. In this arrangement, the passengers can now proceed straight to the shortest queue while mulling on their way down the escalator.

The transparency concept between the landside area and the airside area also allows visual contact between the greeters and the arriving passengers before the latter clear customs. The greeters at the landside area are able to view the arriving passengers walking to the immigration area (Lim, 2008). The JIT principle of continuous improvement can be noted here.

## Airport Clearance

The airport exercises quality control in baggage handling. For the transfer passengers, the airport can promise a Minimum Connection Time (MCT) of 60 minutes (China Economic Review, 2007). This means that the time taken from the moment when the aircraft is parked in the apron area to baggage transfer will be within 60 minutes. For destination passengers, the first baggage will be on the baggage belt after 12 minutes and the last bag from the same aircraft will reach about 29 minutes from the time when the aircraft parks in the apron area. According to the quality standard stipulated by the airport, this has been achieved 90% of the time.

Currently, the airport has set a new goal of clearing 90% of a planeload of passengers through immigration in 8 minutes, 100% of the passengers through customs in 5 minutes and so on. This means the arriving passengers can now expect more speedy and hassle-free exit from the airport (Air Contracts, 2006).

Terminal 3 has a fully-automated baggage system which also incorporates a high speed inter-terminal baggage transfer system (Airport Technology, 2008). For inter-terminal baggage transfer, the baggage is transported through underground tunnels at a speed of 7 meters per second. This means it will take only about three minutes for a bag to be transported through the underground baggage tunnel between Terminal 2 and Terminal 3 (Lim, 2008). As reflected, the JIT principle of total quality control is exercised in the Baggage Handling System (BHS) to maintain high efficiency and quality in delivering the baggage within the time stipulated.

Along the way to the baggage claim area, a large screen displays the belt of the flight number. After passing through immigration, more screens will show the passengers where to retrieve their baggage. The JIT principle relating to the kanban/pull system is reflected here. Necessary information appears when it is needed. Also, cart trolleys are made available at various locations for the passengers to place their hand carry luggage. More importantly, there are also plenty of cart trolleys on standby at the baggage claim area where the passengers need to transport the main pieces of their luggage.

## Accommodation and Rest Areas

For stranded passengers, especially those who are unable to depart from Singapore because of flight problems or other issues, the airport has allocated rest areas complete with bathroom facilities for their use. For longer delays or accommodation, there is the airport hotel, the Crowne Plaza (see Figure 5.29), which is located just next to Terminal 3. Improvements can thus be seen here along the theme of the JIT concept. The hotel is also equipped with amenities and facilities just like any other major hotels with swimming pool, restaurants, spa lounges, health and fitness club, meeting and conference rooms (Changi Airport, 2008).

Figure 5.29. Main view of the Crowne Plaza.

The Crowne Plaza is also ideal in terms of accessibility. With just three minutes walk to the sky-train station, hotel guests can reach Terminal 1 and Terminal 2 conveniently. In addition, it is easy to travel from the hotel to the Changi Airport Mass Rapid Transit (MRT) station to head to other parts of Singapore. In case of disruptions to the MRT system, hotel guests can also choose alternative transport modes using different routes like buses situated in the basement or taxis at the arrival curb side.

Figure 5.30. Taxis waiting outside Terminal 3.

### Taxis

One very commendable feature of the Changi International Airport is the organized taxi lines. The users can wait inside the air-conditioned terminal for the taxis to arrive. These taxis arrive upon demand, thus displaying the JIT principle relating to the kanban system. The taxis do not crowd together at the arrival curb side. Instead, the taxis queue orderly outside Terminal 3 as can be seen in Figure 5.30.

## SUMMARY FOR ARRIVAL AREA

The findings from the questionnaire survey relating to the applicability of the JIT principles for the arrival area is summarized in Table 5.8.

**Table 5.8. JIT principles applied in arrival procedures**

| Airport features | Applicability of JIT principles | Applicability | | |
|---|---|---|---|---|
| | | High | Neutral | Low |
| Lighting conditions | Total Quality Control* | - | - | - |
| People Mover System (PMS) for public and passengers | Elimination of space, time and energy waste | ✓ | | |
| Route to immigration | Uninterrupted work flow | ✓ | | |
| | Continuous improvement* | - | - | - |
| Baggage claim | Total Quality Control* | - | - | - |
| | Kanban/Pull System* | - | - | - |
| Accommodation and rest areas | Continuous improvement* | - | - | - |
| Taxis | Kanban/Pull System | | ✓ | |

(*Excluded in the questionnaire as this seems to be an evident application of the principle.)

## OTHER AIRPORT ASPECTS

Other airport aspects which merit consideration include service quality, staff awards, airline awards and feedback for improvement.

## Quality of Service

The relevant airport authority takes ownership of all problems encountered by any parties such as the government agencies, airport concessionaires, tenants, users and the passengers. Quality of service is overtly professed to be very important for the Changi International Airport. A large part of good quality comes from the vendors; hence, good supplier relations are crucial for delivering excellent airport service. The relevant airport authority may take action against any tenant, supplier or airline arising from a valid complaint (Lim, 2008). If problems are serious, penalties may be imposed.

## Staff Awards

Rewards are appropriately given to the employees who have put in great effort in their work or have outstanding performance. For example, the relevant airport authority recognizes and rewards the frontline employees in the airport who have provided good service as shown in Figure 5.31. These awards include the Service Personality of the Year Award and the Outstanding Service Provider Award (The Straits Times, 2008). This is to motivate frontline staff and other employees to provide excellent customer service.

(Source: Changi Airport, 2006, 2008)

Figure 5.31. Airport employees commended for excellent customer service in 2008.

Training is also provided not just for the frontline staff but all employees who need to upgrade their skills either managerially or technologically for them to derive future benefits for the airport. Once the employees are equipped with better skills and responsibilities for their work, they will be more involved with tasks directed at improving the airport service.

The relevant airport authority also takes effective maintenance work into serious consideration. These include civil engineering works, mechanical and electrical systems, telecommunication facilities and cleaning services. The relevant airport authority ensures that even the top management staff is involved in maintenance issues and expects them to attend to these issues in a timely manner. This shows employee involvement. Since 2007, the relevant airport authority has been conducting integrated trials relating to passenger flows, the integration of various airport systems, as well as total integration (Singapore Sights, 2008) involving systems and human participation (airport staff and other users). In total, more than 50 trials were planned that involved airport staff and the relevant government agencies.

## Airline Awards

Similarly, the same level of excellence applies for supplier relations in raising the competitiveness of the airport. Airlines that carried the most passengers and cargoes through the Changi International Airport as well as airlines that achieved the highest year-on-year growth in passenger and cargo flows were honored with the Changi Airline Awards (Travel Daily News, 2008). 30 awards were given out under four categories: (a) Top 10 Airlines by Passenger Carriage, (b) Top 10 Airlines by Cargo Carriage, (c) Top 5 Airlines by Growth in Passenger Carriage, and (d) Top 5 Airlines by Growth in Cargo Carriage. With continuous support for the airlines, the Changi International Airport continues to maintain the strong partnership with the stakeholders to ensure that new initiatives introduced are aligned with their needs.

## Feedback for Improvement

The management philosophy takes a holistic approach towards coordinating the interests of the government agencies, airport concessionaires, tenants, suppliers and passengers. This demands meticulous planning and execution to serve the customers (passengers, greeters, well-wishers and casual visitors to the airport) well. With supportive participation from employees, there is quality control through the use of a feedback system. The relevant airport authority shares the feedback from the public with other relevant airport agencies to further improve the facilities and services provided in the airport in all the three terminals. The feedback form covers the services provided by the airport agencies and included the check-in facilities and services, friendliness of airport staff, shopping facilities and so on.

## SUMMARY FOR OTHER AIRPORT ASPECTS

The findings from the questionnaire survey relating to the applicability of the JIT principles for other airport aspects are summarized in Table 5.9.

**Table 5.9. JIT principles applied in other airport aspects**

| Airport features | Applicability of JIT principles | Applicability | | |
|---|---|---|---|---|
| | | High | Neutral | Low |
| Quality of service | Supplier relations* | - | - | - |
| Staff awards | Employees involvement | ✓ | | |
| Airline awards | Supplier relations* | - | - | - |
| Feedback | Quality Control* | - | - | - |

(*Excluded in the questionnaire as this seems to be an evident application of the principle.)

*Chapter 6*

# SURVEY FINDINGS

## RESPONSE RATE

Based on the research aim and objectives, a questionnaire was formulated and pilot tested prior to the full scale survey being carried out. A total of 51 questionnaires were distributed to potential respondents with experience in airport management and operations. A total of 33 questionnaires were completed and returned. 33 questionnaires were initially distributed to employees in Changi Airport International (CAI) and 18 questionnaires to employees in the Civil Aviation Authority of Singapore (CAAS). This was both considered to be researcher-administered survey and self-administered survey because there were times when the research team helped the respondents with their queries on the spot. 29 questionnaires were distributed in hard copies while 22 questionnaires were sent via emails. The latter group was self-administered.

There was a higher response rate from the survey given out in hard copies than in soft copies. For hard copies given out on the spot, 24 sets were returned with completed details. These included three questionnaires that were returned by post. This yields a response rate of about 82.7%. For soft copies sent through emails, only nine respondents completed the questionnaire, giving a response rate of 40.9%. The latter process was also longer which took more than a week as compared to a day spent giving out the hard copies in person by the research team.

It appears logical that the response rate for the hard copy route is about twice that of the soft copy route. This was because with the physical presence of a member of the research team, the respondents felt the urgent need to complete the questionnaire. However, sending the questionnaire through emails can reach out to those respondents who were not present at work on the day of the survey or to those respondents working overseas. The entire questionnaire survey exercise was considered to have obtained a high response rate due to a member of the research team having contacts in both the CAI and the CAAS. In addition, eight respondents returned the questionnaire with their written comments (Appendix 3).

## STATISTICAL TESTS

The questionnaire results were analyzed using the Statistical Package for the Social Sciences (SPSS) Statistics 17.0. Three statistical analyses were conducted using the SPSS. The first analysis is the one-sample t-test. The purpose of using the one-sample t-test is to examine if the sample is representative of the entire population to show reliability of the results.

The second analysis uses the mean to find the average rating that the respondents have given to each of the JIT principles. Through the mean, the respondents' views relating to the applicability of each of the JIT principles to the respective airport feature can be ascertained. The neutral value was set at 3 since the rating ranged from 1 to 5 on the Likert scale. If the average rating is greater than 3, it can be said that the respondents as a whole are more receptive to the principle being applicable to the airport features and vice versa. However, knowing the average rating remains inadequate. There can be extreme cases whereby one respondent rated 5 and another respondent rated 1 for the same JIT principle. The overall mean will then be 3. By only interpreting the mean, differing views might be overlooked. Hence, there is the need to conduct a third analysis.

The third analysis is based on the standard deviation. The standard deviation measures how the data are spread out from the mean value. It describes the dispersion of data on either side of a mean value. A low standard deviation indicates that the data set is clustered around the mean value whereas a high standard deviation indicates that the data is widely spread with significantly higher/lower figures than the mean. With the results from this analysis, the differing views can be identified for discussion.

The final sample size obtained was 33. An option was given to the respondents for them to decide if they should give a rating to each JIT principle in the questionnaire. If the principle is inapplicable in their opinions, they can choose not to rate at all by leaving it blank. Thus, some of the results may have sample size, N less than 33 as will be shown later.

# Analysis of Results

Significant results from the survey findings are presented and discussed below.

## One-Sample T-Test

In this test, the smaller the p-value, the more significant the result would be. At a confidence level of 95%, the p-values are statistically significant if it is $\leq 0.05$. This means that the sample used is representative of the population and that the results can be relied upon for evaluation. The more significant data for p-values are shown in Table 6.1. Looking under the column of sig. (2-tailed) p-value, the only outlier is the decentralized security screening feature which is the only one with a p-value of 0.186 which is higher than 0.05.

This airport feature had elicited some comments from the respondents and has shown extreme ratings from 1 to 5. Some concerns relating to the system are that there were other issues that outweigh the principle of merely eliminating space waste. These include the inefficiency caused, long queues for multiple gates, huge amount of duplication of screening equipment installed and others.

**Table 6.1. Results of one-sample t-test**

| JIT principles as applied in Terminal 3 | One-sample t-test Test value = 3 | | | | 95% confidence interval | |
|---|---|---|---|---|---|---|
| | t | df | Sig. (2-tailed) p-value | Mean Diff. | Lower | Upper |
| Cable-net Wall_ Energy Waste | 2.7 | 31 | 0.011 | 0.59 | 0.15 | 1.04 |
| Transparent Pavilion_ Kanban | 2.6 | 29 | 0.013 | 0.53 | 0.12 | 0.95 |
| Flight Information Display System_ Continuous Improvement | 2.7 | 31 | 0.012 | 0.59 | 0.14 | 1.05 |
| Decentralized Security_ Space Waste | 1.4 | 31 | 0.186 | 0.31 | 0.16 | 0.78 |

This outcome appears to arise because the respondents have considered an overall evaluation of the system.

There were also a few respondents with the requisite JIT knowledge who commented that the principle in question is applicable to a large extent (mode of rating is 4). Nevertheless, this airport feature is highly debatable and there appears to be room for improvement in future systems. This issue will be discussed again later with more details under recommendations. The remaining SPSS results can be found in Appendix 4.

## Mean Values

Both the positive and less positive results from the survey findings are presented and discussed below.

### *Positive Results for Future Use*

It is observed that the tabulated results (ranked according to mean values) attached in Appendix 5 tended towards the higher rating of 5. Numerous ratings were in the range of 4 and 5 which pushed the mean value to over 4 for many of the airport features. To some extent, this may suggest that the principles work well in most of the airport features. These techniques are related to increasing quality, productivity and other developments in the airport. Hence, the more applicable the JIT principles, the better it is for advancing future airport developments. In this context, only mean values of above 4.2 will be discussed.

The highest mean value of 4.36 (in Table 6.2) comes from the feature, carts/trolleys whereby uninterrupted work flow is the JIT principle concerned. Having a mean value of 3 means no respondent thinks that the principle is inapplicable. At the same time, the standard deviation is only 0.78 which suggests that the ratings of the respondents are similar to a large extent. This could be an indication that the findings presented in the earlier chapter have been reflected accurately. Placing the cart trolleys beside the entrance doors would make a good practice for continuation in future airport planning as it is seen to have minimal interruption to the route path of the passengers.

Other features with high mean values include the double façade doors, smart roof, pinnacles, conveyor belt, passenger flow paths and staff awards (see Table 6.2). Results for the smart roof and pinnacles are similar with relatively high mean values of 4.26.

**Table 6.2. Data of high rankings**

| Ranking | Airport Features | N | Min. | Max. | Mean | Std. Deviation |
|---|---|---|---|---|---|---|
| 1 | Carts _Uninterrupted Work Flow | 33 | 3.00 | 5.00 | 4.36 | 0.78 |
| 2 | Smart Roof _Energy Waste | 33 | 2.00 | 5.00 | 4.27 | 0.94 |
| 3 | Pinnacles _Space and Energy Waste | 32 | 2.00 | 5.00 | 4.25 | 0.80 |
| 4 | Passenger Flow Paths _Continuous Improvement | 33 | 3.00 | 5.00 | 4.21 | 0.74 |
| 5 | Staff Awards _Employee Involvement | 29 | 3.00 | 5.00 | 4.21 | 0.82 |

Space and energy savings for the pinnacles were justified with prior knowledge. According to one of the designers for the air-conditioning and mechanical ventilation system in Terminal 3, there is significant reduction in energy consumption and thus huge cost savings arising from the use of the pinnacles.

The justification also works well for the passenger flow paths. The high mean value may be due to the layout of the shops. The arrangement is such that the passengers will need to pass by 50% of all the shops. Not all passengers will make a purchase but at least, half the shops have a fair chance of being browsed through and patronized by the passengers, unlike previously. One respondent commented that directing the passengers to go through shops will help increase the airport revenue. Therefore, an improvement has been made. Out of the 33 respondents, four thought that there is no continuous improvement. This may be because of crowd control. With only one main path, in the event of a high volume in the flight schedule at any one time, it may be uncomfortable for the passengers to squeeze through the crowd. Thus, this may explain why the four respondents did not rate this feature highly.

***Less Positive Results for Improvements***

As summarized in Table 6.3, the opinions fluctuated from 1 to 5 for the elements shown, which also accounted for the high standard deviations. The low mean value of 3.31 is for the elimination of space waste for the security screening system.

**Table 6.3. Data of low rankings**

| Ranking | Airport Features | N | Min. | Max. | Mean | Std. Deviation |
|---|---|---|---|---|---|---|
| 28 | Cable-net Wall _Energy Waste | 32 | 1.00 | 5.00 | 3.59 | 1.24 |
| 29 | Flight Information Display System _Continuous Improvement | 32 | 1.00 | 5.00 | 3.59 | 1.27 |
| 30 | Transparent Pavilion _Kanban | 30 | 1.00 | 5.00 | 3.53 | 1.11 |
| 31 | Decentralized Security _Space Waste | 32 | 1.00 | 5.00 | 3.31 | 1.31 |

As noted earlier, the saving of space does not appear to be the main concern of the airport workforce who was involved with the survey. The focus was on security, efficiency and cost. For security purposes, it is necessary to have checks at the individual or combined gates. The respondents would have rated this highly for this purpose. However, although space is saved, this may sometimes create long queues at the common entrance, which can bring down the efficiency level. Cost is also considered. In general, Terminal 3 uses decentralized screening for each individual gate holding room except for the combined gate holding rooms. There appears to be duplication of security screening equipment installed to take care of good safety measures.

It was observed that all the principles that have to do with the kanban system were ranked relatively lower with the highest ranking at 19$^{th}$ place among the 31 attributes. The reasons for this could be that either the respondents have only a vague idea of what the kanban system does or the application has been interpreted wrongly.

However, for the kanban system used in the transparent pavilion, three responses were returned with no rating and have a relatively low average rating of 3.53. There could be a few reasons for this result. Firstly, the photograph in the questionnaire might have been misunderstood by some of the respondents. The research team had meant the entire terminal which adopted the use of glass but the respondents might only relate to the specific area shown in the photograph. Some respondents may therefore think the principle is inapplicable. The low rating might be due to that the extensive use of glass creates confusion to the people's sense of direction instead of helping them. There was a child, it was noted for example, who knocked into the glass but had wanted to go over to the airside at the sky-train station area.

For the three missing ratings, the reason could be that the usage of glass was intended to produce solely the effect of clarity and not to guide people for direction. In addition, the uninterrupted work flow principle for the transparent pavilion may lead to confusion for some respondents as it does not seem to apply to the photograph. Hence, the results for these two principles need not be considered.

Looking at the mean values may not be good enough. These did not highlight two aspects. One relates to those who did not give any rating (see Table 6.4) and how much the data deviates from the mean (please see next section). The double façade entrance doors and cable-net wall do not have high ratings. Perhaps the respondents do not see why the double door is better than a single door to contain the indoor cool air as a form of quality control. It could also be that in reality, having double doors do not help in containing the indoor cool air. For the cable-net wall, the reason could be similar to that of the transparent pavilion because using a glass wall may not allow the users to orientate themselves easily. Furthermore, there can be insufficient factors to contribute towards the glass wall for total quality control. Fritted glass panels used to circumvent the glare problem is possibly not a substantial factor to provide for quality control.

**Table 6.4. Lowest sample size, N**

| Ranking | Airport Features | N | Min. | Max. | Mean |
|---|---|---|---|---|---|
| 17 | Cable-net Wall _Total Quality Control | 27 | 2.00 | 5.00 | 3.96 |
| 21 | Double Façade Doors _ Total Quality Control | 28 | 2.00 | 5.00 | 3.82 |

## Standard Deviations

For the sample standard deviations shown in Table 6.5, the larger results show wider differential views of the same principle. The standard deviation of the security system has the furthest dispersion from the mean and whose possible causes were already discussed earlier. Next, the Flight Information Display System (FIDS) is subjected to the judgment of the users. One respondent commented that the FIDS in Terminal 2 also have individual TV clusters to supplement the consolidated ones. Therefore, the respondent opined that the JIT effect was negated in terms of continuous improvement. Another respondent thought that doing away with the main flight information display board was good because there is saving on space and maintenance.

**Table 6.5. Rankings in terms of standard deviations**

| Ranking | Airport Features | N | Mean | Std. Deviation |
|---|---|---|---|---|
| 12 | Vertical Transport _Total Quality Control | 33 | 4.00 | 0.61 |
| 1 | Carts _Uninterrupted Work Flow | 33 | 4.36 | 0.78 |
| 4 | Passenger Flow Paths _ Continuous Improvement | 33 | 4.21 | 0.74 |
| 28 | Cable-net Wall _Energy Waste | 32 | 3.59 | 1.24 |
| 29 | Flight Information Display System _Continuous Improvement | 32 | 3.59 | 1.27 |
| 31 | Decentralized Security _Space Waste | 32 | 3.31 | 1.31 |

## Evaluation of Results and Recommendations

In this section, the last objective of proposing recommendations for those airport features with rooms for improvement is discussed. The list of recommendations is summarized in Appendix 6 with the suggested actions to be taken. Only some of the more significant actions are discussed below.

### Departure Curb Side

The features to be discussed in this section include the signage and roof, cart trolleys, cable-net wall, double facade entrance doors.

#### *Signage and Roof*

One respondent commented that while the current signage version may have been developed from Terminal 2, nevertheless, there is still room for improvement. The signage can still be modified further because this still looks small when viewed from within a vehicle at a distance. Positive changes made on the signage will directly improve on the kanban system to lead the passengers to the relevant gate entrance. As for the roof, the overall dimensions of the roof can be extended for both its length and width because during rainy days, the driving winds can still rain onto the building glass to affect the passengers.

***Cart Trolleys***

The cart trolleys, ranked first with its high mean value, appear to have actively applied the uninterrupted work flow principle. It is recommended that the airport management should continue to maintain the good work in this area. The functions and operations relating to cart trolleys must continue to maintain efficiency and adequate provisions through reliable suppliers.

***Cable-Net Wall***

There seem to be conflicting results for the cable-net wall. While the wall may be aesthetically pleasing, using the glass façade may increase operating costs and energy consumption in the air-conditioning system and lighting. This may suggest that the materials used in future facades should be reconsidered. The current glass type used in Terminal 3 is based on an improved version used in Terminal 2. If the Changi International Airport is to continue to use glass, better shading devices should be incorporated to reduce the solar heat gain. Alternatively, investments in better glazing systems like double low-emissivity glazing and double glazed façade filled with inert gas can be considered. Otherwise, due to the green house effect of glass, heat tends to be trapped internally that may lead to higher energy consumption for air-conditioning.

More importantly, proper consideration should be given to understand the impact of the choice of façade materials on solar heat gain that may affect energy consumption. Hence, the decision relating to the materials used lies in the design process. Different professionals from different disciplines (constructors, architects, engineers and others) need to discuss this aspect together to attain the best in future airport designs.

***Double-Façade Entrance Doors***

Another feature, the double-façade entrance doors, is also one of the top ranked attributes that can help contribute to eliminating energy waste. This feature was deployed in other terminals and should continue to be used in future designs as well. However, the quality control of the glass doors can be further developed. There may still be some thermal discomfort when someone enters the airport terminal from outdoor or when the air-conditioning is still not well contained within the indoor environment. If properly designed and maintained, this aspect may not be a major concern in the first place. Nevertheless, it remains essential to examine its impact on the users. If the impact is minimal, this feature should still be used in other areas. If the impact is significant, then the changes to be made would depend on the importance of this feature. It should however be acknowledged that the user will pass through the door only in a short period of time.

## Departure/Curb Side Atrium

The major feature to be discussed in this section relates to the "Butterfly" roof.

### *"Butterfly" Roof*

The "Butterfly" roof appears to be well received with mean values above 4. This relates energy savings and in exercising quality control for the amount of natural and artificial lighting admitted into the terminal building. Aesthetically, the "Butterfly" roof is also one of the iconic and architectural features of Terminal 3. Such a roof form should be further encouraged because of both its aesthetic appearance and functionality.

## Check-in Hall

The features to be discussed in this section include the pinnacles, Flight Information Display System (FIDS) and conveyor belt system.

### *Pinnacles – Signage with Air-Conditioning Diffusers*

The air-conditioning and mechanical ventilation (ACMV) system is one of the building elements in the departure hall and the transit mall that supply cold air at the occupancy level. This feature helps to save energy and is built into the interior architecture to save space. Nevertheless, after conducting the site walk-through and attendant measurements, it appears that there is a relatively stronger air draft in the landside such as the departure hall, the shops in level 3 and entrance to the passport control area.

From the on-site measurements, the air velocity measured by the instrument, airflow anemometer, seems to be relatively higher in the areas circled in Figure 6.1. The departure hall area is specifically referred to in Figure 6.2 where the FIDS screens are located between two pinnacles. Strong cold air is felt when the passenger is checking for the information shown on the FIDS screens. This cold draft may cause some thermal discomfort. The same condition seems to be noted in the area in Level 3 between the diffusers situated on top of the shops and the Chinese restaurant when one rides up from the escalator. The condition in this area may, however, appear to be acceptable because most people will not stay at the same spot for too long a time.

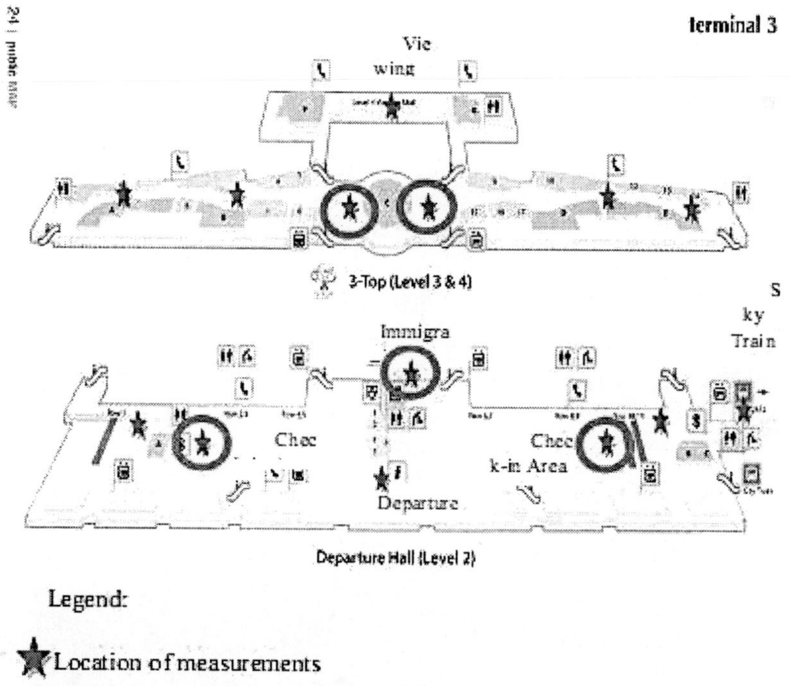


(Source: adapted from Changi Airport Group, 2010)

Figure 6.1. Measurement spots at Levels 3 and 4 in Terminal 3 Departure Hall.

The last area is at the entrance of the departure passport control area (see Figure 6.3) where well-wishers, in sending off their friends/relatives, tend to stay longer at the same spot. In this case, there are diffusers blowing directly at them from both the front and back directions. This spatial layout is only logical because many people tend to gather at the area. However, the air may become too cold for some of them to feel comfortable.

Figure 6.2. Passengers may feel the cold draft when standing between two diffusers.

### *Flight Information Display System (FIDS)*

Some respondents do not appear to see this feature as an improvement because the FIDS in Terminal 2 have additional individual television clusters to supplement the consolidated ones seen in Terminal 3. However, this perception seems to be related more to the considerations of minimized congestion, smooth passenger route path and costs. If there are improvements to be made, functionality and the abovementioned considerations need to be taken into account collectively to derive the best value for the FIDS.

(a)

(b)

Figure 6.3. (a) and (b) Diffusers blowing directly at airport users from front and back directions.

***Conveyor Belt System***

There appears to be scope for the conveyor belt system to be improved further. The angle of the conveyor belt system to receive the passenger luggage had been adjusted to better facilitate baggage movements. However, this does not seem to work that well in reality. There were instances where the baggage does not fall flat to be inducted into the take-away belt. In such instances, the airline employees will still need to get up from their seats to place the baggage flat on the conveyor belt. Elimination of motion (effort) waste is mirrored to a certain extent in such instances. The slope of the angle may perhaps need to be redefined for it to work well as intended. Alternatively, for example, an additional slope for the conveyor belt can be set up at the take-away belt. Enhancing this function can bring the conveyor belt system to deliver better effectiveness, thus reflecting the process of continuous improvement.

## Retail, Restaurants and other Concession Areas

The features to be discusssed in this section include the retail shops and amenities as well as viewing gallery.

***Retail Shops and Amenities***

From the many international accolades won, the Changi International Airport undoubtedly provides one of the best shopping experiences in the world. For example, Changi International Airport has won the Skytrax title as the best airport for leisure amenities. Similarly in 2008, the Changi International Airport was named the "Best Duty-Free Airport in the World" for the 7th consecutive year (Travel Daily News, 2008). The shopping amenities in Terminal 3 are still in keeping with excellent global retail standards. Nevertheless, from on-line reviews posted by the airport users, it is noted that the shops in Terminal 3 were perceived to cater only to the up-scale passengers (Skytrax, 2009). This may well be a deliberate corporate strategy adopted by Terminal 3 to cater more to the luxury segment of the business in order to justify costs as well as to fit into meeting the needs of the luxurious class of travelers. Nevertheless, a better mix to also include the mid-ranged travelers may be considered and recommended for quality control.

Figure 6.4. Viewing gallery in Terminal 3 in the day.

***Viewing Gallery***

In addition to the retail amenities, the viewing gallery is also an area where improvements can be considered. Although there is transparency from using glass for the visitors to view the aircrafts in the day (see Figure 6.4), the glass can cause reflections at night (see Figure 6.5). Consequently, when the visitors look out of the viewing gallery at night, their ability to see the external view may be affected by reflections. It is however acknowledged that there may not be much scenery to view at night in the airfield. The public users may also find it difficult to look from a closer view because there is a void between the glass façade and the viewing area. These issues do not seem to be major ones but can nonetheless provide the opportunity for considering good quality control to further meet user requirements.

Figure 6.5. Viewing gallery in Terminal 3 at night.

## Transit Mall

The features to be discussed in this section include the passenger flow path and design of the shop frontage.

### *Passenger Flow Path*

It appears that commercial operations are secondary to the core functions of the airport. At times, commercial activities may impede the airport operations. On the improvements that can be made to the passenger flow path, one respondent suggested that there should not be only one main path. There can be by-pass paths for passengers, who are in tight flight connection situations, to avoid the crowded commercial areas. While this may be the case, the spatial arrangement should also ensure that these paths do not cause the passengers to miss out on passing by and patronizing most of the shops.

At the same time, the users of the terminals in the Changi International Airport have feedback on the relatively longer pathway to the gates in Terminal 3. This is due in part to the large and generous space provided. This appears to be the case especially for the last few gates at both extreme ends of the terminal building. If the gate is on one side, there may be instances where the travelers are discouraged from shopping at the opposite side of the passenger flow path because of the longer walk required. Thus, this may have an adverse effect on the retail shops.

In addition, there may also be conflicts operationally and commercially. Anxious or late passengers may just want to get to the boarding gates immediately but they nevertheless need to go through the long passenger flow path lined with shops. However, if travelators were to be installed straight from the immigration area onwards, the normal passengers may tend to use the travelators and thereby skip through the entire shopping experience. The spatial layout and arrangement may therefore need to be considered very carefully. This is because the improvements made for the retail area may compromise on some other aspects in the building. Future solutions may need to balance between the two considerations, depending on the strategic priorities of the airport.

### *Design of Shop Frontage*

Some respondents suggested that although the zig-zag frontage is functional, it may not be pleasing to the eye of some passengers. This is because the passengers may be confused by too many shop frontages that is arranged in a zig-zag manner. This aspect may however be subjective in the opinions of some passengers. Nevertheless, since this feedback was provided, there must have been some passengers who felt this way. Further quality control can be considered to enhance this feature.

## Gate Lounges

The key feature to be discussed in this section relates to the decentralized security screening system.

***Decentralized Security Screening System***

Based on the feedback received, there seems to be room for improvement in the decentralized security screening system. The elimination of space waste alone appears insufficient. Apart from attaining high efficiency and good coordination of flight schedules, there is also a need to deploy minimal resources (screening equipment and manpower) as far as possible. The suggestion of implementing mini "centralized" security screening system for combined gate holding rooms may indeed lead to the use of less resource than in using the decentralized system for all individual gate holding rooms. This suggestion is recommended for consideration and implementation with modifications, if appropriate.

The "express" lane concept may also be considered. The "express" lane can allow passengers, who are in a hurry to catch their planes, go through a particular lane more quickly. For example, fifteen minutes before the closing of an aircraft gate, an express lane opens and gives priority to passengers boarding that aircraft to go through the security checks first. However, in the process, security rigor must never be compromised. Alternatively, there can be a "speedy" lane for passengers who have one or no hand held luggage.

Excellent coordination of flight schedules is also necessary so that the timings for different aircrafts under the same gate lounge should not be arranged too close together at any one time. However, it is acknowledged it may be quite difficult at times to estimate or control the timings, especially for connecting flight schedules.

To increase flexibility and efficiency, and subject to availability, another consideration is to provide for a backup gate lounge or to switch to a less crowded gate lounge for security checks. Although the passenger flow path may not be as smooth as a result, it might be better at times when the passengers are really in a hurry to board their planes. Lastly, the security system can be shared by fewer gates although this would mean resources would need to be increased as well. However, larger aircrafts like the Airbus 380 and Boeing 747 should continue to have individual gates for security checks as these planes have to cater to a larger pool of passengers.

## Arrival Area

The features to be discussed in this section include the People Mover System (PMS)/sky-trains and taxis.

### *People Mover System (PMS)/Sky-Trains*

For the PMS, two respondents commented that the issues relating to energy consumption may not be that applicable. They observed that the system may appear to be inefficient in terms of operating expenditure because this had increased from one PMS line to over five PMS lines which have implications for increased energy consumption. However, it is acknowledged that if it is necessary to have these lines for reasons of expediency and timeliness, costs may not be of the most importance concerns. But this issue should be borne in mind with the addition of future airport terminals which may require the use of many more PMS lines for good inter-terminal connectivity.

### *Taxis*

It is a highly commendable practice in the Changi International Airport that the passengers can continue to stay in the air-conditioned area of the terminal building while waiting for the cabs to arrive. However, while this practice may be good for the cab users, this may not appear to be the case for the cab drivers who are waiting in the queue outside. It is quite common to see many cabs queuing for some time outside the terminal building before they can pick up the passengers. Sometimes, when the other terminals require more cabs, the cab drivers are unable to get out of the queue to get to the other terminals in time. This may give rise to inefficient service quality for the cab users and frustration to the cab drivers. For this reason, the road network can consider providing more exit openings in future airport designs and upgrading programs.

## Other Airport Aspects

The key feature to be discussed in this section relates to the staff and airline awards.

### *Staff and Airline Awards*

The staff and airline awards serve as motivation for employee involvement and airline cooperation. These accomplishments can be extended to more categories of employees to be considered for the awards. This will certainly encourage more stakeholders to put in more efforts to maintain and enhance the well-being of the airport operations and management. Considering more categories would mean the need to explore more areas that have excellent performance but were not formally recognized previously.

Alternatively, more could be done to include a greater number of awards/staff benefits under the original categories. The airport budget and management policies may be extended to consider this.

Even though not all stakeholders appear to be aware of the JIT concept, the existing airport operations have actually already incorporated some forms of JIT principles in their designs, planning and management. After having been briefed about the JIT concept, it appears that the respondents had already unknowingly adopted some elements of JIT principles in the work which they do. This is despite the fact that the respondents had never heard of the JIT principles being applied to the spatial layout in airports prior to the explanation. It should also be noted that some of the JIT principles are closely linked to certain airport features. Fulfilling one JIT principle may however, but not necessarily, lead to another feature being undermined. Consequently, it is crucial for the airport to recognize the outcome of what the JIT principles can bring about based on its priorities and given resources.

From the questionnaire survey, a respondent may rate a particular JIT principle and corresponding airport feature very highly on a Likert scale of 5 for applicability. However, this does not necessarily translate to successful implementation. Nonetheless, the JIT principles must be emphasized for continuous improvements. By the same token, having low applicability between a particular JIT principle and corresponding airport feature may suggest that things can be done better, leading to greater success for the airport when implemented in the future.

*Chapter 7*

# CONCLUSION

## BACKGROUND

While the primary aim of this book is to examine how the JIT principles may be extended for managing the spatial layout in airports, it should be acknowledged that the application of the JIT principles in the built environment is not entirely new. The JIT principles have been applied for managing prefabrication yards, the deliveries of precast concrete components, the production and deliveries of ready-mixed concrete, site layout management and the design of ramp up light factories. These are discussed briefly below as a precursor before the conclusion of this study relating to JIT management for airports is presented.

## JIT CONCEPT IN PREFABRICATION YARDS

A study on the application of JIT principles in prefabrication yards was completed by Low and Chan (1996). Low and Chan's (1996) empirical findings and their subsequent analysis in the light of the JIT concept have indicated that there is potential for applying the JIT principles to improve productivity in the prefabrication sector.

From the JIT viewpoint, it was pointed out that for a proposed physical factory layout in a prefabrication yard to operate efficiently and effectively, the following two facilities must be provided for:

1. Overhead gantry cranes of sufficient capacity and operating length to cover the entire physical factory layout. This would ensure total

lifting services to be provided for all operations within the factory layout, and
2. Access ways and roads of adequate sizes and number for use by workmen and machinery (fork-lifts and lorry-trailers).

Similarly, a process layout can be proposed for implementation by the prefabricators of building components. The important point is to separate out all the various production sub-process activities in a flow-line in sequential order. In this way, the existing problem of utilizing the same factory space for both concrete casting and finishing treatment works (being prevalent among prefabricators of building components) can be resolved.

It should also be noted that the main drawback would be the high initial capital costs incurred in setting-up such large factory spaces in order to conform with the JIT principles. Nonetheless, this should be critically weighed against the increased productivity and profitability in the long term (Low and Chan, 1996).

## JIT Deliveries of Precast Components

According to the study completed by Low and Choong (2001), there is excellent potential for applying JIT principles for the deliveries of precast concrete components to building sites. However, there appears to be a lack of mutual confidence between the precasters and the contractors to effect the JIT system of delivery. Contractors lamented the occurrences of late deliveries, while the precasters attributed the blame to inaccurate demand schedules, slow revision and updates of changes, as well as last-minute demands by the contractors. The onus to rectify this mistrust in order to implement the JIT concept seems to rest with the contractors.

Specific and precise contractual agreements can serve as a good launch pad to harness mutual confidence, i.e., contracted confidence. Responsibilities, expectations, shouldering of costs, and other minute details must be clearly spelled out in the contract and understood by both parties. Discipline in strictly executing the contract is the next operational task. This involves adherence to the demand schedule provided by the contractor, with minimal changes in the structural stage, and reliable support by the precasters.

The contractors are, by and large, driven and motivated by price and have generally overlooked the larger picture of total costs. Only a minority of the respondents in Low and Choong's (2001) study showed an interest in JIT deliveries as a cost-saving management system. Conversely, the precasters appear to be more prepared to provide logistical support for JIT deliveries. In this context, reimbursement of costs or sharing of the savings can help to motivate the precasters to commit to JIT deliveries.

## JIT Concept in the Ready-Mixed Concrete Industry

Another study by Low and Wu (2005) found ready application of the JIT principles in the ready-mixed concrete (RMC) industry. Low and Wu (2005) conducted a survey covering twenty and fifteen RMC suppliers in Chongqing, China and Singapore respectively. The study indicated that all the RMC suppliers were using the demand pull system to manage their batching and delivery of RMC. The study also suggests that both the traditional Economic Order Quantities (EOQ) system and the JIT system were adopted to manage the procurement of raw materials in the RMC industry.

RMC suppliers may adopt JIT purchasing to mobilize aggregates, sand and concrete admixtures from the vendors simultaneously. They can then operate faster, eliminate waste, achieve continuous improvement, improve customer service, and build organizational competitiveness.

Low and Wu's (2005) study showed that the procurement of raw materials in RMC batching plants may be made through the traditional EOQ (or non-JIT) approach, the JIT-purchasing approach or a combination of both approaches. Depending on their operational constraints, some batching plants may continue to regard the traditional EOQ procurement system to be more viable than the JIT purchasing system for procuring raw materials, and vice-versa. Nevertheless, in the final analysis, Low and Wu's (2005) study confirmed that regardless of physical locations, it would appear that RMC suppliers universally used the demand pull system in the batching and delivering of RMC to sites.

## JIT Concept in Site Layout Management

Low and Mok (1999) conducted a study to examine how the JIT principles have been applied explicitly to an actual private condominium site to improve its site layout effectively. The study concluded that it is possible to apply the various JIT principles for site layout to improve productivity and that the main difficulties in the implementation of the JIT concept are primarily people-related problems. The JIT concept was seen as a philosophy, a method of achieving high productivity and quality with little or no wastage. Without a cooperative workforce, its implementation would reap none of the benefits experienced in the manufacturing industry. Low and Mok's (1999) study also highlighted four groups of people whom careful consideration, education and training must be given before the JIT principles can be implemented successfully in the construction industry. These four groups of people are the clients/consultants, suppliers, subcontractors and workers.

## JIT Design for Ramp-up Light Factories

The concept of ramp-up light factories is becoming more popular among developers in recent years. These factories not only maximize land utilization but also help to resolve problems found in the traditional multi-storey factories by providing direct access to every unit using the vehicle ramp and internal driveway.

However, since the introduction of these ramp-up factories, little study had been made to understand its impact on raising the operational productivity of the tenant's facilities in terms of cost, time and quality. With increasing success of the JIT principles as applied in other areas (such as in manufacturing) and in the construction stage, there is a need to examine how the operational productivity of the tenant's facilities can be met through the application of the JIT principles for improving the facilities design of these ramp-up factories.

The needs of the tenant's operational facilities in terms of space and layout of the factory were examined by Low and Show (2008). This study appears to suggest that the needs of the operational facilities that were considered important by the architects, who designed these factories, were similar to the tenants' requirements. This seems to suggest that the architects have, to some extent, considered the tenants' requirements for the needs of the operational facilities during the design stage.

The architect's design in relation to the JIT principles in meeting the needs of the tenants' facilities was also examined. It was found that for each of the JIT principles, the tenants were generally quite satisfied with all the facilities design features highlighted.

The study completed by Low and Show (2008) suggests that the respondents were, on the whole, quite satisfied with the following facilities design features that were provided:

1. Doorstep loading/unloading bay
2. Vehicular ramp and internal driveway
3. Wide frontage
4. Clear floor to ceiling height

The study by Low and Show (2008) also suggests that the two JIT principles of "Elimination of waste" and "Uninterrupted work flow" were more applicable to the design layout of facilities. With the introduction of vehicular ramp, internal driveway and doorstep loading/unloading bays, direct access to the tenants' doorstep is provided. This would eventually help to reduce waiting time, double handling of goods during transportation and motion waste. In addition, it also provides for smooth deliveries to every unit with less damage to the quality of the goods.

In order to successfully apply the JIT principles to improving the facilities design of ramp-up light factories, the architect must, first of all, fully understand the JIT principles and how these are closely inter-related principles. With further training on planning their internal facilities layout, the tenants would be able to reap the benefits of such an improved design model. From the pedagogical viewpoint, it is therefore recommended that architecture schools take into consideration the concepts presented by Low and Show (2008) to alert designers of the need to incorporate relevant JIT principles into the designs of the facilities to further enhance the operational requirements of the building users downstream.

Based on the outcome of the study completed by Low and Show (2008), it appears that the tenants generally agreed that the application of the JIT principles would effectively help to meet the needs of their operational facilities and improve their productivity. However, the origin of the JIT principles was from the manufacturing industry. Hence, in applying and improving the facilities design layout of ramp-up light factories, it should be acknowledged that some modifications to the JIT principles may be necessary to attain appropriateness.

## IMPLICATIONS

Three major misconceptions over how the JIT concept operates are apparent from the completed studies described above. Firstly, the JIT concept is often construed as an inventory control system. Although the implementation structure of the JIT concept will include the control of inventory, this is not its major and only function. Secondly, the JIT concept is also misinterpreted as a method used by the material function to push inventory back to the supplier, thereby forcing the supplier to carry the customer's inventory. The third misconception also suggests that JIT systems are essentially quality control programs. This is, however, also misleading because the JIT concept is an operational philosophy and product quality as such must always come from the manufacturing processes. These misconceptions should first be overcome before the JIT concept can be considered for implementation. In addition to correct interpretation and understanding, the implementation of JIT systems is often flawed with numerous pitfalls. Some of these pitfalls include:

1. Failure to obtain commitment from top management
2. Inadequate education programs
3. Inadequate external assistance
4. Underestimating the task of implementation
5. Inappropriate time scale for implementation
6. Failure to integrate the process and control operations
7. Lack of software and appropriate integration with the financial and accounting system
8. Rush towards maintaining supplier links
9. Failure to appreciate the JIT concept as a long-term on-going process

Nonetheless, despite these misconceptions and pitfalls in implementation, there are attractive benefits for organizations to adopt JIT systems for their operations. The major benefits would include cost savings, enhanced quality and productivity performance.

While the benefits of improved productivity are obvious in JIT based manufacturing systems, some modifications would be required to the JIT concept for it to be applied meaningfully in the construction industry. Construction activities are not only complex. The industry is also fragmented with different professionals, suppliers and contractual arrangements. Nevertheless, where construction is concerned, the JIT concept appears gainfully relevant for the building materials industry with process or repetitive manufacturing processes. Although it may be more difficult for the JIT concept to be applied in one-off building projects, this may not be so where industrialization, which makes use of prefabricated components, is concerned in the construction industry. The use of standard components in prefabrication, which mirrors the repetitive processes found in manufacturing, therefore offers an environment which is conducive for the application of the JIT concept (Low and Chan, 1996). In addition, studies have shown that the JIT concept can be applied successfully for the deliveries of precast concrete components to building sites (Low and Choong, 2001), in the ready-mixed concrete industry (Low and Wu, 2005) and in site layout management (Low and Mok, 1999).

In summary, as the application of the JIT concept in the construction industry is an area of work which requires more study, the following two recommendations are suggested for further research:

1. Establish the differences in performance (in terms of time, costs, quality and the environment) between similar building projects which have implemented the JIT principles and those which have not.
2. One of the key objectives of the JIT concept is to reduce inventory costs. The reduction in inventory costs, with its consequential effects on cash flow, is recommended for further research.

## CONCLUSION

The lessons learned from the above studies are useful to support and conclude the current study relating to the application of the JIT concept for airport management. Quite apart from its application in other areas of the built environment, the JIT concept was specifically examined in this study in the context of the Changi International Airport through its planning processes and existing operations. This appears to be successful because the study was able to highlight the strengths as well as areas for potential improvements in the airport through the application of the seven JIT principles.

The design principles observed in Terminal 3 of the Changi International Airport were achieved drawing on past experiences and based on careful thought given by the designers. Continuous improvement in future airport designs can be made possible with further development of the recommendations and advancements using the latest technology for airport operations. This will bring about further benefits to the airport. An increase in productivity can lead to efficiency and cost savings. Consequently, profits can be enhanced for the airport which in turn brings about better bonuses, incentives and rewards for everyone, especially the employees. In the process, it is hoped that the employees will continue to work harder and to explore more innovative ways to serve the airport. It is therefore important for the whole organization to work together towards the common goal of providing the best for the customers consistently.

After reviewing the applicability of the JIT principles within the context of the airport features, it is noted that these principles were indeed relevant and useful for analyzing the various operational aspects found in the Changi International Airport. However, the presence of some limitations should be noted.

## LIMITATIONS

The one limitation associated with this study is that the analysis was based on various sources of information. In this context, there could be other sources of data not covered that may highlight other aspects of the airport features and JIT principles in greater details. In addition, the research method adopted as well as the target groups of respondents in the study posed their own sets of limitations as well.

In so far as the research method is concerned, the questionnaire itself did not document the full list of JIT principles that may be of relevance to the airport features. The intention was to keep the survey as short as possible by omitting those principles that may not be evaluated initially to be that impactful for the study. For some airport features, when considering the principles involved, certain factors may not be accounted for within the overall context of these features due to a lack of the relevant knowledge involved.

Choices in the questionnaire may also lead to perceptibly less reliable data sets. For example, the scale of “4” and “5” may mean different standards to different subjects interpreting the questions. After consulting some respondents and analyzing the data, it was found that some respondents were still unsure of certain JIT principles and their applicability despite the explanations provided to them at the onset. Some had unintentionally compared the airport features with those either in other terminals or other airports. There were also two respondents who circled two ratings on the same row for the same principle, leaving the row above/below blank. All these led to inaccuracies in the results generated, which should be accounted for when interpreting the findings.

It was also found to be useful to know from the respondents why a certain rating was given or no rating was given. However, only a few respondents have commented on some airport features. This was because there might be other reasons which influenced their decisions in the ratings. However, given the exhaustive list of 31 features, it would be difficult to ask them to comment on each and every one of these features.

The survey was conducted only with the staff from the airport. Another consideration should be placed on the expectations and requirements of the passengers. The ultimate goal of all airport features is really for the benefit of the passengers. Hence, it is important to know how they feel about these airport features in order to know whether these features have achieved what they were intended to be designed and built for in the first place.

## FURTHER RESEARCH

To learn from the perspective of the passengers, the questionnaire survey can be administered to travelers who have passed through the immigration and operational procedures in both Terminal 2 and Terminal 3. A sample of the questionnaire is shown in Appendix 7. The questionnaire is based on the same airport features. But the questions are phrased in an indirect way. For example, to find out whether there is improvement in the signage being larger and clearer for the users to see, the question relates to what extent is the clarity of the signage from afar. The passengers may likewise rate from a scale of 1 to 5 for consistency.

The survey should also include the optional comments to collate additional information on the airport features. The survey itself should be accompanied by photographs with explanations as shown earlier in Appendix 2 because some of the scenarios are best explained using photographs. Lastly, as far as possible, the survey should be a researcher-administered questionnaire exercise. This is to help the respondents with any queries which they may have relating to the questions asked.

Despite some of the limitations faced, the three objectives of the research had been achieved. With changes made, the original JIT principles can indeed be applied in the context of airport terminals. These principles are able to bring out the strengths and weaknesses of each of the airport elements as highlighted in the findings and evaluation, thus fulfilling the first two objectives of this study. After understanding these elements well enough, the weaknesses can then be highlighted and recommendations proposed for future airport design and planning. In this manner, the last objective has been accomplished.

# Appendix 1. Sample of Questionnaire with Explanations

## A. Brief Introduction

The objective of this questionnaire is to examine how the Just-In-Time (JIT) principles have been applied in Terminal 3. The study hopes to evaluate how the JIT principles can further be applied in future airport development projects for enhancement. The questions asked relate mainly to Terminal 3 but you may need to make reference to Terminal 2 when answering some questions.

## B. JIT Principles

The seven JIT principles are explained below :

1) Elimination of waste – any activity that does not add value to a product or process; any wastage in terms of time, effort and money, etc. These can include :

a) Waste of time
b) Waste of motion (effort)
c) Waste of space
d) Waste of energy

2) Kanban/Pull System – produce only when needed; supply at the right time.

3) Uninterrupted work flow – continuous, synchronized and smooth processes, with minimum disruption to work processes. These can include :

a) Focused Factory – grouping of a single family of items/processes arranged together in the layout of Terminal 3.
b) Group Technology – grouping the components of a family of similar parts in a continuous layout that leads from one process/procedure to another.
c) Design for Simplification and Automation – simplify processses before introducing automation.

4) Total Quality Control – ensure quality at work is achieved.
5) Employee involvement – a workforce that is well-trained, participative and plays a part in decision making in specific matters.
6) Supplier relations – suppliers providing quality products in the right quantities at the right time to suit customer needs so as to maintain good airport service quality.
7) Continuous improvement – any improvement made in Terminal 3 but not found in Terminal 2.

# APPENDIX 2. SUPPORTING PHOTOGRAPHS FOR THE QUESTIONNAIRE

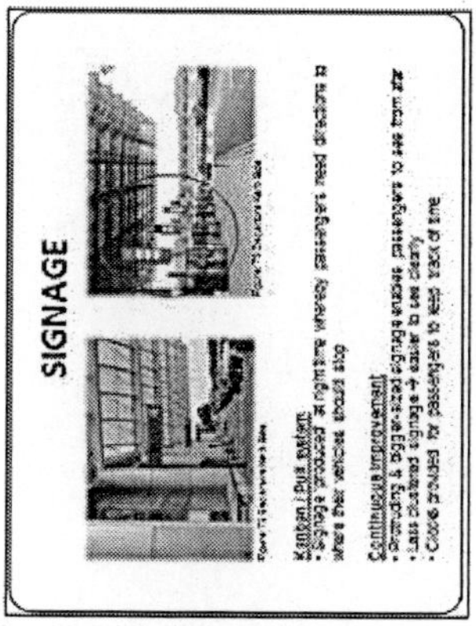

## CABLE-NET WALL

**Total Quality Control**
- Clarity makes it easy for users to orientate themselves
- Fritted glass panels used to counteract glare problem

**Elimination of Energy Waste**
- Improved glass façade facilitate maximal light transmission but minimal direct heat from sun
→ more daylighting allowed into building but reduced artificial lighting

## DOUBLE-FAÇADE ENTRANCE DOOR

**Total Quality Control**
- Air-conditioning contained within indoor environment
- Gradual temperature change for users entering terminal
→ Cool and comfortable

**Elimination of Energy Waste**
- Higher temperature can be set since cool air will not escape easily as compared to single automated door

## DEPARTURE ATRIUM

## "BUTTERFLY" ROOF

**Total Quality Control**
- Skylights automatically adjust to sunlight
→ allows right amount of natural light
- Prevent excessive glare & heat
- Tube lights to keep light intensity same regardless of sun condition
→ providing soothing ambience all day

**Elimination of Energy Waste**
- Low-E coating applied to glass
→ admit 70% visible daylight and 35% thermal heat

## TRANSPARENT PAVILION

Figure: T3 Departure Check-in Hall

**Kanban / Pull System**
- Extensive use of glass induces users to subconsciously know in advance where to head to for the next procedure

**Uninterrupted Work Flow– Group Technology (GT)**
- Clarity of spatial layout allows continuation of process leading from one to another

## ESCALATORS, LIFTS & TRAVELLATORS

Figure: T3 Departure Kerb Side

**Elimination of Time & Energy Waste**
- If arrived at wrong level, users can use these equipment (near entrance) to switch level fast → minimal time wasted
- Motion sensors only activated when in use → minimal energy wasted

**Total Quality Control**
- Single storey concept for departure & arrival halls → No need for change in level
- In case of breakdowns, alternative equipment can be used

## CHECK-IN HALL

## PINNACLES - SIGNAGE WITH AIR-CONDITIONING DIFFUSERS

Figure: A typical pinnacle in T3 Check-in hall

**Elimination of Space & Energy Waste**
- Air-conditioning units are cleverly integrated into the structure, pinnacles together with signage (no.11 as shown) → saves space
- Cool air now supplied from pinnacles at occupant level → saves energy

# FLIGHT INFORMATION DISPLAY BOARD (FIDS)

Figure: FIDS in T2 Departure Hall

Figure: FIDS in T3 Departure Hall

**Continuous Improvement**
- Travelers can now go straight to the respective check-in counters instead of walking to centralized FIDS board (shown in T2)
- Resolves congestion issue found in T2 previously & smooth walk path

# CHECK-IN LAYOUT

Figure: T3 Check-in Hall

**Total Quality Control**
- 1.4m² area per passenger to achieve comfort & level of service A (IATA)
- Distance passengers has to walk from kerb side to closest check-in point kept minimum

**Uninterrupted Work Flow– Focused Factory**
- Grouping of several airlines companies for check-in purposes
- Sharing of check-in counters by airlines under the same alliance umbrella

# BAGGAGE HANDLING SYSTEM

Figure: T3 Conveyor Belt System

**Continuous Improvement**
- Induction section (circled in figure) tilted to facilitate baggage to fall flat onto conveyor belt smoothly
- Every 2 counters in T3 share usage of 1 odd-sized belt instead of 4 counters sharing 1 in T2
- Automated early bag storage facility

# CONVEYOR BELT SYSTEM

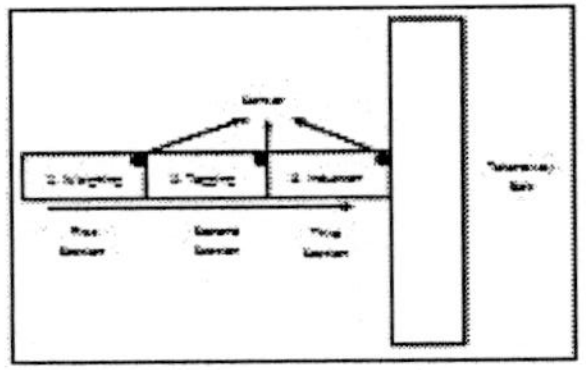

Figure: T3 Conveyor Belt System and its operations

**Elimination of Waste of Motion (Effort)**
- Previously, weighing & tagging done together in same section
→For several baggage, airline employee has to move baggage as only one baggage can be inducted at one time
- Now, weighing & tagging done separately (as shown) as tagging section can hold several baggage at a time
→ Employee need not shift baggage

## RETAIL, RESTAURANTS AND OTHER CONCESSION AREAS

## RETAIL SHOPS AND AMENITIES

Figure: T3 retail shops

**Kanban / Pull System**

- Plan for most shops to be inside transit mall to "pull" passengers to go through immigration early
→ Avoid large surge of passengers & for revenue purpose

**Total Quality Control**

- Enjoy more than 100 retail & over 40 F&B outlets with wide ranging services from established local & international brands

## IMMIGRATION AREA

## LIGHTINGS CONDITION

Figure: Outside T3 Departure/Immigration Area

**Total Quality Control**

- Lightings inside immigration area gradually become dimmer as compared to those used in departure hall
→ Create relaxed environment for travelers to enjoy a wonderful shopping experience in transit mall

# TRANSIT MALL

# PASSENGER FLOW PATHS

Figure: T3 Transit Mall Layout

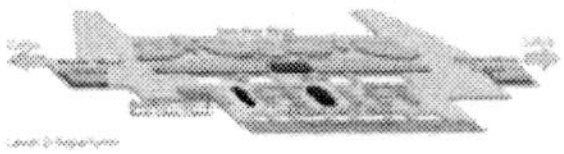

Figure: T3 Departure Hall and Transit Mall Layout

**Continuous improvement**

- Primary (main) path unclear
→ Passengers tend to miss out parts of shopping experience if taken secondary paths due to hierarchy of pathways in T2

- One obvious pathway (either left or right) to gate in T3
→ Either way will pass through about half of retail outlets

# DESIGN OF SHOP FRONTAGE

Figure: T3 Transit Mall Shops

**Total Quality Control**

- Shops designed in zig-zag manner (curved rather than straight, flat)
→ Passengers can see the kinds of shops ahead as they walk
→ Enhance shops' attractiveness

# GATE LOUNGES

## DECENTRALIZED SECURITY SCREENING

Figure: T3 Departure Gate Lounges

**Elimination of Space Waste**
• Instead of having individual gate lounge, several lounges space can be combined into one
→ Same goes for security screening checks and equipment at one area

## ARRIVAL

## ROUTE TO ARRIVAL IMMIGRATION

Figure: T3 Arrival Immigration Area

**Uninterrupted Work Flow**
• Passengers have to go down one floor via escalators
→ To locate the shortest queue from view above
→ Smooth work flow without stopping

## PEOPLE MOVER SYSTEM (PMS)

Figure: T3 PMS station

**Elimination of Time Waiting, Energy and Space Waste**

• One PMS for both public users and travelers instead of two

• Doors only opened on one side for each carriage; public & travelers' side

→ Saves space of building additional track, energy consumption of two sky trains & time waiting as well

## CABS

Figure: T3 outside where cabs are parked

**Kanban / Pull System**

- Cabs in queue

→ Only allowed to leave & fetch passengers when demanded

## OTHERS

## STAFF AWARDS

Figure: Changi staff commended for customer service in 2004

**Employee Involvement**

- Presented awards such "Service Personality of the Year" & "Outstanding Service Providers"

→ motivate frontline staff and others to provide good customer service

# APPENDIX 3. COMMENTS INDICATED BY RESPONDENTS IN THE QUESTIONNAIRE

## C. QUESTIONNAIRE

For each of the airport elements, please rate 1 to 5 for the applicability of the relevant JIT principles to the element with 1 being the least applicable and 5 being the most applicable. However, you can choose not to rate (by leaving the row blank) if in your opinion, the JIT principles are not applicable in that case.

| AIRPORT ELEMENTS | APPLICABLITY OF JIT PRINCIPLES | RATING (1 to 5) | | | | |
|---|---|---|---|---|---|---|
| Example: | | | | | | |
| Sheltered car parks | Continuous Improvement (compared with T1 in this example)<br>Thinking process:<br>T1 uses open air car parks<br>T2 uses sheltered car parks<br>Shelters users from bad weather conditions and more lots created<br>Improvement made (vs. T1)<br>In my opinion, this JIT principle is applicable so I rate it as 5. | 1 | 2 | 3 | 4 | 5 |
| (1) T3 DEPARTURE CURB SIDE | | | | | | |
| Signage | Kanban/Pull System | 1 | 2 | 3 | 4 | 5 |
| | Continuous Improvement (vs. T2) | 1 | 2 | 3 | 4 | 5 |

| | | | | | | |
|---|---|---|---|---|---|---|
| Roof | Total Quality Control | 1 | 2 | 3 | 4 | 5 |
| Carts/Trolleys | Uninterrupted Work Flow | 1 | 2 | 3 | 4 | 5 |
| | Supplier Relations | 1 | 2 | 3 | 4 | 5 |
| Cable-net Wall | Elimination of Energy Waste | 1 | 2 | 3 | 4 | 5 |
| | Total Quality Control | 1 | 2 | 3 | 4 | 5 |
| Double-façade Entrance Doors | Elimination of Energy Waste | 1 | 2 | 3 | 4 | 5 |
| | Total Quality Control | 1 | 2 | 3 | 4 | 5 |
| Comments (you may wish to add in any input):<br>Signage could be improved. Still small when viewed from vehicles approaching from a distance. JIT principles evident, but not optimized.<br>Roof not of sufficient width as winds do blow the rain onto the building glass.<br>JIT principles seen clearly on trolley functions and operations.<br>Not clear where fritted glass fits into a JIT system. | | | | | | |
| (2) DEPARTURE/CURB SIDE ATRIUM | | | | | | |
| "Butterfly" Roof | Elimination of Energy Waste | 1 | 2 | 3 | 4 | 5 |
| | Total Quality Control | 1 | 2 | 3 | 4 | 5 |
| Transparent Pavilion | Kanban/Pull System | 1 | 2 | 3 | 4 | 5 |
| | Uninterrupted Work Flow - GT | 1 | 2 | 3 | 4 | 5 |
| Lifts, Escalators and Travellators | Elimination of Time and Energy Waste | 1 | 2 | 3 | 4 | 5 |
| | Total Quality Control | 1 | 2 | 3 | 4 | 5 |
| Comments (you may wish to add in any input):<br>Roof and Pavilion provide good example of JIT principles.<br>Don't fully agree on Elimination of Time and Energy Waste although they are sufficient in numbers. | | | | | | |
| (3) CHECK-IN HALL | | | | | | |
| Pinnacles - signage with Air-conditioning Diffusers | Elimination of Space and Energy Waste | 1 | 2 | 3 | 4 | 5 |
| Flight Display Information Board (FIDS) | Continuous Improvement (vs. T2) | 1 | 2 | 3 | 4 | 5 |
| Check-in Layout | Total Quality Control | 1 | 2 | 3 | 4 | 5 |
| | Uninterrupted Work Flow | 1 | 2 | 3 | 4 | 5 |
| Conveyor Belt System | Continuous Improvement (vs. T2) | 1 | 2 | 3 | 4 | 5 |
| | Elimination of Waste of Motion | 1 | 2 | 3 | 4 | 5 |

Comments (you may wish to add in any input):
Air-con Pinnacles provide for good examples of JIT principles.
FIDS in T2 also have individual TV clusters to supplement the consolidated ones, therefore, JIT effects here negated in terms of Continuous Improvement.
Uninterrupted Work Flow for Check-In Layout could be improved, but T3 still has one of the best examples as compared to regional airports.

(4) RETAIL, RESTAURANTS AND OTHER CONCESSION AREAS

| | | | | | | |
|---|---|---|---|---|---|---|
| Retail Shops and Amenities | Kanban/Pull System | 1 | 2 | 3 | 4 | 5 |
| | Total Quality Control | 1 | 2 | 3 | 4 | 5 |

Comments (you may wish to add in any input):
Commercial operations are secondary to the core functions of the airport. At times, commercial activities impede the airport operations. Pull factor not necessary although it would be welcomed to some passengers.
Quality can be improved, as T3 shops cater to the up-scale passengers. Should be a better mix for the mid-range passengers as well.

(5) IMMIGRATION AREA

| | | | | | | |
|---|---|---|---|---|---|---|
| Lighting condition | Total Quality Control | 1 | 2 | 3 | 4 | 5 |

Comments (you may wish to add in any input):

(6) TRANSIT MALL

| | | | | | | |
|---|---|---|---|---|---|---|
| Passenger flow paths | Continuous Improvement (vs. T2) | 1 | 2 | 3 | 4 | 5 |
| Design of shop frontage (zig-zag manner) | Total Quality Control | 1 | 2 | 3 | 4 | 5 |

Comments (you may wish to add in any input):
Passenger Flow Paths should not be only one main path. Bypass paths needed for passengers in tight connection situations to avoid crowded commercial areas.
Zig-zag Frontage although functional, not as pleasing to the eye of some passengers as an over-presence of Zig-Zag Frontage can lead to confusion.

(7) GATE LOUNGES

| | | | | | | |
|---|---|---|---|---|---|---|
| Decentralized Security Screening | Elimination of Space Waste | 1 | 2 | 3 | 4 | 5 |

Comments (you may wish to add in any input):
Excellent use of JIT principles.

| (8) ARRIVAL | | | | | | |
|---|---|---|---|---|---|---|
| Route to Immigration | Uninterrupted Work Flow | 1 | 2 | 3 | 4 | 5 |
| People Mover System (PMS) for public and passengers | Elimination of Space, Time and Energy Waste | 1 | 2 | 3 | 4 | 5 |
| Cabs | Kanban/Pull System | 1 | 2 | 3 | 4 | 5 |
| Comments (you may wish to add in any input):<br>Supply for cabs not properly evident in JIT system, although demand principles are addressed. | | | | | | |
| (9) OTHERS | | | | | | |
| Staff Awards | Employees Involvement | 1 | 2 | 3 | 4 | 5 |

## C. QUESTIONNAIRE

For each of the airport elements, please rate 1 to 5 for the applicability of the relevant JIT principles to the elements with 1 being the least applicable and 5 being the most applicable. However, you can choose not to rate (by leaving the row blank) if in your opinion, the JIT principles are not applicable in that case.

| (1) T3 DEPARTURE CURB SIDE | | | | | | |
|---|---|---|---|---|---|---|
| Signage | Kanban/Pull System | 1 | 2 | 3 | 4 | 5 |
| | Continuous Improvement (vs. T2) | 1 | 2 | 3 | 4 | 5 |
| Roof | Total Quality Control | 1 | 2 | 3 | 4 | 5 |
| Carts/Trolleys | Uninterrupted Work Flow | 1 | 2 | 3 | 4 | 5 |
| | Supplier Relations | 1 | 2 | 3 | 4 | 5 |
| Cable-net Wall | Elimination of Energy Waste | 1 | 2 | 3 | 4 | 5 |
| | Total Quality Control | 1 | 2 | 3 | 4 | 5 |
| Double-façade Entrance Doors | Elimination of Energy Waste | 1 | 2 | 3 | 4 | 5 |
| | Total Quality Control | 1 | 2 | 3 | 4 | 5 |
| Comments (you may wish to add in any input): | | | | | | |
| (2) DEPARTURE/CURB SIDE ATRIUM | | | | | | |
| "Butterfly" Roof | Elimination of Energy Waste | 1 | 2 | 3 | 4 | 5 |
| | Total Quality Control | 1 | 2 | 3 | 4 | 5 |

| | | | | | | |
|---|---|---|---|---|---|---|
| Transparent Pavilion | Kanban / Pull System | 1 | 2 | 3 | 4 | 5 |
| | Uninterrupted Work Flow - GT | 1 | 2 | 3 | 4 | 5 |
| Lifts, Escalators and Travellators | Elimination of Time and Energy Waste | 1 | 2 | 3 | 4 | 5 |
| | Total Quality Control | 1 | 2 | 3 | 4 | 5 |
| Comments (you may wish to add in any input): | | | | | | |
| (3) CHECK-IN HALL | | | | | | |
| Pinnacles - signage with Air-conditioning Diffusers | Elimination of Space and Energy Waste | 1 | 2 | 3 | 4 | 5 |
| Flight Display Information Board (FIDS) | Continuous Improvement (vs. T2) | 1 | 2 | 3 | 4 | 5 |
| Check-in Layout | Total Quality Control | 1 | 2 | 3 | 4 | 5 |
| | Uninterrupted Work Flow | 1 | 2 | 3 | 4 | 5 |
| Conveyor Belt System | Continuous Improvement (vs. T2) | 1 | 2 | 3 | 4 | 5 |
| | Elimination of Waste of Motion | 1 | 2 | 3 | 4 | 5 |
| Comments (you may wish to add in any input):<br>There are FIDS located around T2 as well. | | | | | | |
| (4) RETAIL, RESTAURANTS AND OTHER CONCESSION AREAS | | | | | | |
| Retail Shops and Amenities | Kanban/Pull System | 1 | 2 | 3 | 4 | 5 |
| | Total Quality Control | 1 | 2 | 3 | 4 | 5 |
| Comments (you may wish to add in any input): | | | | | | |
| (5) IMMIGRATION AREA | | | | | | |
| Lighting condition | Total Quality Control | 1 | 2 | 3 | 4 | 5 |
| Comments (you may wish to add in any input):<br>Should be the use of artificial lighting versus natural light at work. | | | | | | |
| (6) TRANSIT MALL | | | | | | |
| Passenger flow paths | Continuous Improvement (vs. T2) | 1 | 2 | 3 | 4 | 5 |
| Design of shop frontage (zig-zag manner) | Total Quality Control | 1 | 2 | 3 | 4 | 5 |
| Comments (you may wish to add in any input): | | | | | | |

| | | | | | | |
|---|---|---|---|---|---|---|
| | | | | | | |
| (7) GATE LOUNGES | | | | | | |
| Decentralized Security Screening | Elimination of Space Waste | 1 | 2 | 3 | 4 | 5 |
| Comments (you may wish to add in any input):<br>Example shown is that of a combined gate holding room with centralized screening for the particular few gates. | | | | | | |
| (8) ARRIVAL | | | | | | |
| Route to Immigration | Uninterrupted Work Flow | 1 | 2 | 3 | 4 | 5 |
| People Mover System (PMS) for public and passengers | Elimination of Space, Time and Energy Waste | 1 | 2 | 3 | 4 | 5 |
| Cabs | Kanban/Pull System | 1 | 2 | 3 | 4 | 5 |
| Comments (you may wish to add in any input):<br>Regarding PMS, save energy consumption and time waiting not applicable. | | | | | | |
| (9) OTHERS | | | | | | |
| Staff Awards | Employees Involvement | 1 | 2 | 3 | 4 | 5 |

# Appendix 4. SPSS Results of One-Sample T-Test

| SPSS statistics | | | | |
|---|---|---|---|---|
| JIT Principles as Applied in T3 | N | Mean | Std. Deviation | Std. Error Mean |
| Signage _ Kanban | 30 | 3.9000 | .92289 | .16850 |
| Signage _Continuous Improvement | 33 | 4.0606 | .99810 | .17375 |
| Roof _Total Quality Control | 33 | 3.9697 | 1.07485 | .18711 |
| Carts _Uninterrupted Work Flow | 33 | 4.3636 | .78335 | .13636 |
| Carts _Supplier Relations | 32 | 3.7813 | 1.23744 | .21875 |
| Cable-net Wall _Energy Waste | 32 | 3.5938 | 1.24069 | .21933 |
| Cable-net Wall _Total Quality Control | 27 | 3.9630 | 1.12597 | .21669 |
| Double Façade Doors _Energy Waste | 32 | 4.1875 | .82060 | .14506 |
| Double Façade Doors _ Total Quality Control | 28 | 3.8214 | .86297 | .16309 |
| Smart Roof _Energy Waste | 33 | 4.2727 | .94448 | .16441 |
| Smart Roof _ Total Quality Control | 30 | 4.0000 | 1.11417 | .20342 |
| Transparent Pavilion _Kanban | 30 | 3.5333 | 1.10589 | .20191 |
| Transparent Pavilion _ Uninterrupted Work Flow _Group Technology | 32 | 3.6875 | 1.06066 | .18750 |
| Vertical Transport _Time W. _Energy W. | 33 | 3.7879 | .96039 | .16718 |
| Vertical Transport_ Total Quality Control | 33 | 4.0000 | .61237 | .10660 |
| Pinnacles _Space Waste _Energy Waste | 32 | 4.2500 | .80322 | .14199 |
| FIDS_ Continuous Improvement | 32 | 3.5938 | 1.26642 | .22387 |
| Check-in Layout_ Total Quality Control | 30 | 4.0667 | .82768 | .15111 |
| Check-in Layout _ Uninterrupted Work Flow | 32 | 4.0938 | .89296 | .15785 |

| | | | | |
|---|---|---|---|---|
| Conveyor Belt_ Continuous Improvement | 32 | 4.1875 | .59229 | .10470 |
| Conveyor Belt _Motion Waste | 32 | 3.9063 | 1.05828 | .18708 |
| Retail Shops Amenities _Kanban | 33 | 3.7576 | 1.14647 | .19957 |
| Retail Shops Amenities _TQC | 33 | 4.0000 | 1.08972 | .18970 |
| Lighting Condition _ Total Quality Control | 32 | 3.7188 | .77186 | .13645 |
| Passenger Flow Paths_ Continuous Improvement | 33 | 4.2121 | .73983 | .12879 |
| Shop Frontage Design_ Total Quality Control | 33 | 3.7879 | .99240 | .17275 |
| Decentralised Security _Space Waste | 32 | 3.3125 | 1.30600 | .23087 |
| Immigration Route _UWF | 33 | 4.0000 | .86603 | .15076 |
| PMS _Space W. _Time W. _Energy W. | 33 | 4.0303 | 1.01504 | .17670 |
| Cabs _Kanban | 33 | 3.8788 | .92728 | .16142 |
| Staff Awards _Employees Involvement | 29 | 4.2069 | .81851 | .15199 |

| JIT Principles as Applied in T3 | **One-sample t-test** Test Value = 3 | | | | 95% Confidence Interval | |
|---|---|---|---|---|---|---|
| | t | df | Sig. (2-tailed) | Mean Diff. | Lower | Upper |
| Signage _ Kanban | 5.341 | 29 | .000 | .90000 | .5554 | 1.2446 |
| Signage _Continuous Imp. | 6.104 | 32 | .000 | 1.06061 | .7067 | 1.4145 |
| Roof _Total Quality Control | 5.183 | 32 | .000 | .96970 | .5886 | 1.3508 |
| Carts _Uninterrupted Work Flow | 10.000 | 32 | .000 | 1.36364 | 1.0859 | 1.6414 |
| Carts _Supplier Relations | 3.571 | 31 | .001 | .78125 | .3351 | 1.2274 |
| Cable-net Wall _Energy Waste | 2.707 | 31 | .011 | .59375 | .1464 | 1.0411 |
| Cable-net Wall _Total Quality C. | 4.444 | 26 | .000 | .96296 | .5175 | 1.4084 |
| Double Façade Doors_Energy W. | 8.186 | 31 | .000 | 1.18750 | .8916 | 1.4834 |
| Double Façade Doors _ Total Quality Control | 5.037 | 27 | .000 | .82143 | .4868 | 1.1561 |
| Smart Roof _Energy Waste | 7.741 | 32 | .000 | 1.27273 | .9378 | 1.6076 |
| Smart Roof _ Total Quality C. | 4.916 | 29 | .000 | 1.00000 | .5840 | 1.4160 |
| Transparent Pavilion _Kanban | 2.641 | 29 | .013 | .53333 | .1204 | .9463 |
| Transparent Pavilion _ UWF | 3.667 | 31 | .001 | .68750 | .3051 | 1.0699 |

| | | | | | | |
|---|---|---|---|---|---|---|
| _Group Technology | | | | | | |
| Vertical Transport _Time W. _Energy W. | 4.713 | 32 | .000 | .78788 | .4473 | 1.1284 |
| Vertical Transport_ Total Quality Control | 9.381 | 32 | .000 | 1.00000 | .7829 | 1.2171 |
| Pinnacles _Space W. _Energy W. | 8.803 | 31 | .000 | 1.25000 | .9604 | 1.5396 |
| FIDS_ Continuous Improvement | 2.652 | 31 | .012 | .59375 | .1372 | 1.0503 |
| Check-in Layout_ Total Quality C. | 7.059 | 29 | .000 | 1.06667 | .7576 | 1.3757 |
| Check-in Layout _ UWF | 6.929 | 31 | .000 | 1.09375 | .7718 | 1.4157 |
| Conveyor Belt_ Continuous Imp. | 11.342 | 31 | .000 | 1.18750 | .9740 | 1.4010 |
| Conveyor Belt _Motion Waste | 4.844 | 31 | .000 | .90625 | .5247 | 1.2878 |
| Retail Shops Amenities _Kanban | 3.796 | 32 | .001 | .75758 | .3511 | 1.1641 |
| Retail Shops Amenities _TQC | 5.272 | 32 | .000 | 1.00000 | .6136 | 1.3864 |
| Lighting Condition _ Total Quality Control | 5.268 | 31 | .000 | .71875 | .4405 | .9970 |
| Passenger Flow Paths_ Continuous Improvement | 9.412 | 32 | .000 | 1.21212 | .9498 | 1.4745 |
| Shop Frontage Design_ Total Quality Control | 4.561 | 32 | .000 | .78788 | .4360 | 1.1398 |
| Decentralised Security _Space W. | 1.354 | 31 | .186 | .31250 | -.1584 | .7834 |
| Immigration Route _UWF | 6.633 | 32 | .000 | 1.00000 | .6929 | 1.3071 |
| PMS _Space W. _Time W. _Energy W. | 5.831 | 32 | .000 | 1.03030 | .6704 | 1.3902 |
| Cabs _Kanban | 5.444 | 32 | .000 | .87879 | .5500 | 1.2076 |
| Staff Awards _Employees Inv. | 7.940 | 28 | .000 | 1.20690 | .8956 | 1.5182 |

# APPENDIX 5. RANKINGS OF MEANS WITH STANDARD DEVIATIONS

| Rank | Airport Features | N | Min | Max | Mean | Std Deviation |
|---|---|---|---|---|---|---|
| 1 | Carts _UWF | 33 | 3.00 | 5.00 | 4.3636 | 0.78335 |
| 2 | Smart Roof _Energy Waste | 33 | 2.00 | 5.00 | 4.2727 | 0.94448 |
| 3 | Pinnacles _Space W. _Energy W. | 32 | 2.00 | 5.00 | 4.2500 | 0.80322 |
| 4 | Passenger Flow Paths _CI | 33 | 3.00 | 5.00 | 4.2121 | 0.73983 |
| 5 | Staff Awards_ Employees Inv. | 29 | 3.00 | 5.00 | 4.2069 | 0.81851 |
| 6 | Conveyor Belt _CI | 32 | 3.00 | 5.00 | 4.1875 | 0.59229 |
| 7 | Double Façade Doors _Energy W. | 32 | 3.00 | 5.00 | 4.1875 | 0.82060 |
| 8 | Check-in Layout _UWF | 32 | 2.00 | 5.00 | 4.0938 | 0.89296 |
| 9 | Check-in Layout _Total Quality C. | 30 | 2.00 | 5.00 | 4.0667 | 0.82768 |
| 10 | Signage _Continuous Imp. | 33 | 2.00 | 5.00 | 4.0606 | 0.99810 |
| 11 | PMS _SpaceW_TimeW_EnergyW | 33 | 1.00 | 5.00 | 4.0303 | 1.01504 |
| 12 | Vertical Transport _Total QC | 33 | 2.00 | 5.00 | 4.0000 | 0.61237 |
| 13 | Immigration Route _UWF | 33 | 1.00 | 5.00 | 4.0000 | 0.86603 |
| 14 | Retail Shops Amenities _Total QC | 33 | 2.00 | 5.00 | 4.0000 | 1.08972 |
| 15 | Smart Roof _ Total Quality C. | 30 | 1.00 | 5.00 | 4.0000 | 1.11417 |
| 16 | Roof_ Total Quality C. | 33 | 2.00 | 5.00 | 3.9697 | 1.07485 |

| | | | | | | |
|---|---|---|---|---|---|---|
| 17 | Cable-net Wall_ Total Quality C. | 27 | 2.00 | 5.00 | 3.9630 | 1.12597 |
| 18 | Conveyor Belt _Motion W. | 32 | 1.00 | 5.00 | 3.9063 | 1.05828 |
| 19 | Signage _Kanban | 30 | 2.00 | 5.00 | 3.9000 | 0.92289 |
| 20 | Cabs _Kanban | 33 | 2.00 | 5.00 | 3.8788 | 0.92728 |
| 21 | Double Façade Doors_Total QC | 28 | 2.00 | 5.00 | 3.8214 | 0.86297 |
| 22 | Vertical Transport _TimeW_EnergyW | 33 | 1.00 | 5.00 | 3.7879 | 0.96039 |
| 23 | Shop Frontage Design _Total QC | 33 | 1.00 | 5.00 | 3.7879 | 0.99240 |
| 24 | Carts _Supplier Relations | 32 | 1.00 | 5.00 | 3.7813 | 1.23744 |
| 25 | Retail Shops Amenities _Kanban | 33 | 2.00 | 5.00 | 3.7576 | 1.14647 |
| 26 | Lighting Condition _Total QC | 32 | 3.00 | 5.00 | 3.7188 | 0.77186 |
| 27 | Transparent Pavilion _UWF_GT | 32 | 1.00 | 5.00 | 3.6875 | 1.06066 |
| 28 | Cable-net Wall _Energy W. | 32 | 1.00 | 5.00 | 3.5938 | 1.24069 |
| 29 | FIDS_ Continuous Improvement | 32 | 1.00 | 5.00 | 3.5938 | 1.26642 |
| 30 | Transparent Pavilion _Kanban | 30 | 1.00 | 5.00 | 3.5333 | 1.10589 |
| 31 | Decentralised Security _Space W. | 32 | 1.00 | 5.00 | 3.3125 | 1.30600 |

# APPENDIX 6. SUMMARY OF RECOMMENDATIONS

| AIRPORT FEATURES | APPLICABLITY OF JIT PRINCIPLES | APPLICABLITY | | | SUGGESTED ACTIONS |
|---|---|---|---|---|---|
| | | High | Neut | Low | |
| (1) T3 DEPARTURE CURB SIDE | | | | | |
| Signage | Kanban/Pull System | | ✓ | | - |
| | Continuous Improvement | ✓ | | | Improve |
| Roof | Total Quality Control | | ✓ | | Improve |
| Ramp | Continuous Improvement* | - | - | - | - |
| Carts/ Trolleys | Uninterrupted Work Flow | ✓ | | | Maintain |
| | Supplier Relations | | ✓ | | Maintain |
| Cable-net Wall | Elimination of Energy Waste | | | ✓ | Improve |
| | Total Quality Control | | ✓ | | Improve |
| Double-façade Entrance Doors | Elimination of Energy Waste | ✓ | | | Maintain |
| | Total Quality Control | | ✓ | | Improve |
| (2) DEPARTURE/CURB SIDE ATRIUM | | | | | |
| "Butterfly" Roof | Elimination of Energy Waste | ✓ | | | Maintain |
| | Total Quality Control | ✓ | | | Maintain |
| Transparent Pavilion | Kanban/Pull System | | | ✓ | Improve |
| | Uninterrupted Work Flow - GT | | | ✓ | Improve |
| Lifts, Escalators and Travelators | Elimination of Time and Energy Waste | | ✓ | | - |
| | Total Quality Control | ✓ | | | Maintain |
| Signage | Kanban/Pull System** | - | - | - | Maintain |

| | | | | | |
|---|---|---|---|---|---|
| (3) CHECK-IN HALL | | | | | |
| Pinnacles | Elimination of Space and Energy Waste | ✓ | | | Improve |
| Flight Information Display System (FIDS) | Uninterrupted Work Flow* | - | - | - | - |
| | Continuous Improvement | | | ✓ | Improve |
| Check-in Layout | Uninterrupted Work Flow – Focused factory and Design for Simplification and Automation | ✓ | | | Maintain |
| | Quality Control | ✓ | | | Maintain |
| Conveyor Belt System | Elimination of Waste of Motion | | ✓ | | Improve |
| | Uninterrupted Work Flow* | - | - | - | - |
| | Continuous Improvement | ✓ | | | Improve |
| AIRPORT FEATURES | APPLICABLITY OF JIT PRINCIPLES | APPLICABLITY | | | SUGGESTED ACTIONS |
| | | High | Neut | Low | |
| (4) RETAIL, RESTAURANTS AND OTHER CONCESSION AREAS | | | | | |
| Retail Shops and Amenities | Kanban/Pull System | | ✓ | | - |
| | Total Quality Control | ✓ | | | Improve |
| Viewing gallery | Quality Control* | - | - | - | Improve |
| (5) IMMIGRATION AREA | | | | | |
| Signage | Kanban/Pull System** | - | - | - | - |
| Lightings condition | Quality Control | | ✓ | | Maintain |
| (6) TRANSIT MALL | | | | | |
| Passenger flow paths | Supplier Relations* | - | - | - | - |
| | Continuous Improvement | ✓ | | | Improve Further |
| | Kanban/Pull System** | - | - | - | Maintain |
| Design of shop frontage | Total Quality Control | | ✓ | | Improve |
| | Uninterrupted Work Flow** - Group Technology | - | - | - | Maintain |

| | | | | | |
|---|---|---|---|---|---|
| (zig-zag manner) | | | | | |
| Rest areas | Quality Control* | - | - | - | Maintain |
| (7) GATE LOUNGES | | | | | |
| Pathway to gate lounges | Continuous Improvement* | - | - | - | Maintain |
| Decentrali-zed Security Screening | Elimination of Space Waste | | | ✓ | Improve |
| Gate Holding rooms | Total Quality Control* | - | - | - | Maintain |
| | Continuous Improvement* | - | - | - | Maintain |
| (8) ARRIVAL | | | | | |
| Lighting Condition | Total Quality Control** | - | - | - | Maintain |
| People Mover System (PMS) for public and passengers | Elimination of Space, Time and Energy Waste | ✓ | | | Maintain |
| Route to Immigration | Uninterrupted Work Flow | ✓ | | | Maintain |
| | Continuous Improvement* | | | | |
| Baggage Claim | Total Quality Control* | - | - | - | Maintain |
| | Kanban/Pull System** | - | - | - | Maintain |
| Accommo-dation and Rest Areas | Continuous Improvement* | - | - | - | Maintain |
| Taxis | Kanban/Pull System | | ✓ | | Improve |
| AIRPORT FEATURES | APPLICABLITY OF JIT PRINCIPLES | APPLICABLITY | | | SUGGESTED ACTIONS |
| | | High | Neut | Low | |
| (9) OTHERS | | | | | |
| Quality of Service | Supplier Relations* | - | - | - | - |
| Staff Awards | Employees Involvement | ✓ | | | Maintain |
| Airline | Supplier Relations** | - | - | - | - |

| | | | | | |
|---|---|---|---|---|---|
| Awards | | | | | |
| Feedback | Quality Control* | - | - | - | Improve |

* Excluded in the questionnaire - evident application of principle

** Excluded in the questionnaire - similar to that as previously described

# Appendix 7. Sample Questionnaire for Passengers in Terminal 3

| | | | | | |
|---|---|---|---|---|---|
| 1. What was the purpose of you going to Terminal 3? (Please tick) | | | | | |
| (a) I was a departing passenger from T3 [Please proceed to Qn. (2)] | | | | | |
| (b) I was an arriving passenger to T3 [Please proceed to Qn. (16)] | | | | | |
| (c) I was both a departing and arriving passenger in T3 [Please proceed to Qn.(2)] | | | | | |
| | | | | | |
| 2. How did you get to the airport? (Please tick) | | | | | |
| (a) Bus | | | | | |
| (b) Mass Rapid Transit (MRT) train | | | | | |
| (c) Taxi/Drop off at Departure curb side | | | | | |
| (d) Private vehicles and parked in the car parks | | | | | |
| Please answer from Qn. (3) onwards if you had dropped off at the departure curb side. If you did not, please proceed to Qn. (6). Please rate 1 to 5 (least helpful to most helpful) for the following questions. If you cannot recall, please leave the row blank. | | | | | |
| | | | | | |
| TERMINAL 3 DEPARTURE CURB SIDE | | | | | |
| 3. Recalling from memory, how clear do you think the signage was from afar? | 1 | 2 | 3 | 4 | 5 |
| 4. In your opinion then, how would you rate the sufficiency of the carts/trolleys? | 1 | 2 | 3 | 4 | 5 |
| 5. To what extent do you think the air-conditioning was contained well in the internal environment when you first stepped into the terminal? | 1 | 2 | 3 | 4 | 5 |
| Comments (you may wish to provide any feedback): | | | | | |
| DEPARTURE /CURB SIDE ATRIUM | | | | | |
| 6. Throughout the stay in the departure hall, to what extent do you think the lighting was consistent? | 1 | 2 | 3 | 4 | 5 |

| | | | | | |
|---|---|---|---|---|---|
| 7. How clear do you think the transparent pavilion help you to navigate your way through the airport along the way? | 1 | 2 | 3 | 4 | 5 |
| 8. To what extent do you think was the accessibility to vertical transportation like escalators, lifts and travelators? | 1 | 2 | 3 | 4 | 5 |
| Comments (you may wish to provide any feedback): | | | | | |
| CHECK-IN HALL | | | | | |
| 9. How comfortable were you with the air-conditioning environment at the check-in area? | 1 | 2 | 3 | 4 | 5 |
| 10. For the following items, how would you rate them for airport service quality? | | | | | |
| (a) Space allocated per passenger with/without using trolley | 1 | 2 | 3 | 4 | 5 |
| (b) Distance walked from drop-off points to check-in hall | 1 | 2 | 3 | 4 | 5 |
| (c) Queue waiting time | 1 | 2 | 3 | 4 | 5 |
| Comments (you may wish to provide any feedback): | | | | | |
| RETAIL, RESTAURANTS AND OTHER CONCESSION AREAS | | | | | |
| 11. To what extent would you rate the quality of facilities and amenities in terms of variety and brands? | 1 | 2 | 3 | 4 | 5 |
| Comments (you may wish to provide any feedback): | | | | | |
| IMMIGRATION AREA | | | | | |
| 12. How comfortable were you with the lighting conditions beyond the passport control area as compared to the departure hall? | 1 | 2 | 3 | 4 | 5 |
| Comments (you may wish to provide any feedback): | | | | | |
| TRANSIT MALL | | | | | |
| 13. To what extent does the shop design allow you to see its content from afar? | 1 | 2 | 3 | 4 | 5 |
| Comments (you may wish to provide any feedback): | | | | | |
| GATE LOUNGES | | | | | |
| 14. What kind of security system did you go through? (Please tick) | | | | | |
| (a) Individual security screening for each gate holding room | | | | | |
| (b) Combined security screening for 2 gate holding rooms | | | | | |
| (c) Combined security screening for more than 2 gate holding rooms | | | | | |
| 15. How satisfactory was the waiting time for security checks? | 1 | 2 | 3 | 4 | 5 |
| Comments (you may wish to provide any feedback): | | | | | |
| You do not need to answer this section if you have not experienced the arrival procedures in Terminal 3. | | | | | |

| ARRIVAL | | | | | |
|---|---|---|---|---|---|
| 16. To what extent do you think it is good to be one level up when walking towards the immigration area in order to identify the shortest queue? | 1 | 2 | 3 | 4 | 5 |
| 17. Are you satisfied with waiting time to claim your baggage? | 1 | 2 | 3 | 4 | 5 |
| 18. If you have taken a cab from the airport, how satisfactory were you with the waiting time and quality of the waiting experience? | 1 | 2 | 3 | 4 | 5 |
| Comments (you may wish to provide any feedback): | | | | | |
| OTHERS | | | | | |
| 19. In your opinion, how satisfactory was the overall performance of the airport? | 1 | 2 | 3 | 4 | 5 |
| Comments (you may wish to provide any feedback): | | | | | |

# REFERENCES


Air Contracts (2006) *East Asia and Pacific Creating Benchmark in Airport Experience*, Retrieved on 27th September 2008 from: http://www.lnnnews.com/magazines.

Airliners (2009) *The Boeing 747*, Retrieved on 10th January 2009 from: http://www.airliners.net/aircraft-data/stats.main?id=100.

Airport Technology (2008) *Changi Airport (SIN/WSSS), Singapore*, Retrieved on 20th September 2008 from: http://www.airport-technology.com/projects/changi/.

Ansari A. and Modarress B. (1990) *Just-In-Time Purchasing,* The Free Press, New York.

APICS (2008) *Advancing Productivity, Innovation and Competitive Success,* The Association for Operations Management, Retrieved on 16th January 2009 from: http://www.apics.org/apics/SearchResults.aspx?query.

Balakrishnan R. (2003) *The Toyota Production System - A Case Study of Creativity and Innovation in Automotive Engineering,* Retrieved on 11th February 2009 from: http://www.bluecanopy.com/pdf/The_Toyota_Production_System.pdf.

Bowen D. and Youngdahl W. (1998) Lean service: in defence of a production-line approach, *International Journal of Service Industry Management,* 9(3):207-225.

Broadbent M. (2008) *The World's Best Airports*, Retrieved on 9th January 2009 from: http://www.airport-int.com/news/2008/02/28/the-worlds-best-airports.

Broyles D. et al. (2005) *Just-In-Time Inventory Management Strategy and Lean Manufacturing,* Retrieved on 27th December 2008 from:

http://www.academicmind.com/unpublishedpapers/business/operationsmanagement/2005-04-000aaf-just-in-time-inventory-management.

Changi Airport (2006) *Changi Airport Singapore,* Civil Aviation Authority of Singapore, Retrieved on 3rd July 2008 from: http://www.changiairport.com/changi/en/index.html.

Changi Airport (2008) *Changi Airport Singapore Terminal 3,* Civil Aviation Authority of Singapore, Retrieved on 26th July 2008 from: http://www.changiairport.ebizintel.com/index.html.

Changi Airport Group (2010) *Airport Guide. Singapore Changi Airport,* Changi Airport Group (Singapore) Private Limited.

Cheng T.C.E. and Podolsky S. (1996) *Just-In-Time Manufacturing: An Introduction,* 2nd ed. Chapman and Hall, London.

China Economic Review (2007) *CX shortens connect time with Air China,* Retrieved on 27th September 2008 from: http://www.chinaeconomicreview. com/hotels.

Flight Memory (2008) *Airport Singapore (Singapore) – Change*, Retrieved on 22nd January 2009 from: http://www.flightmemory.com/encyclopedia/Singapore_Changi_Airport.html.

Foo S. M. (2009) Of 14 best airport awards this year, Changi has bagged 12, *The Straits Times,* 25 June, p.A31.

Geoghegan T. (2008) *What makes a good airport?* Retrieved on 7th January 2009 from: http://news.bbc.co.uk/2/hi/uk_news/magazine/ 7290432.stm.

Goddard W. E. (1986) *Just-In-Time*, Oliver Wight Ltd. Publications, Inc.

Greenberg D. (2002) *Just-In-Time Inventory System Proves Vulnerable to Labor Strife,* Los Angeles Business Journal, Retrieved on 27th December 2008 from: http://www.findarticles.com/p/articles/mi_m5072/is_40_24/ai_ 930 09975.

Hay E. J. (1988). *The Just-In-Time Breakthrough. Implementing the New Manufacturing Basics*, John Wiley and Sons, Inc.

IATA (2004) *Airport Development Reference Manual*, 9th ed., International Air Transport Association, Montreal.

Johnson C. (2001) *Just In Time - Industry Trend or Event? The Industry Standard*, Standard Media International, Retrieved on 27th December 2008 from: http://www.findarticles.com/p/articles/mi_m0HWW/is_8_4/ai_ 71324 386.

Joseph F. (2008) *Too LeJIT to quit*, Wordpress, Retrieved on 30th January 2009 from: http://committeeofpublicsafety.wordpress.com/2008/10/18/too-lejit-to-quit/.

Karlsson C. and Norr C. (1994) Total Effectiveness of Just-In-Time System, *International Journal of Operations and Production Management,* 14(3):46-65.

Kaur K. (2009) MM Lee: Let's prepare Changi to stay ahead, *The Straits Times,* 2 July, p.A3.

Kaur K. (2009a) No guarantees for Changi, *The Straits Times,* 15 July, p.A2.

Kumar R. (2005) *Research methodology,* Sage Publication Ltd, London.

Liang D. (2009) Changi needs more buzz, *Today,* 15 July, p.4.

Lim L. Y. and Low S. P. (1992) *Just-in-Time Productivity for Construction*, SNP Publishers Pte Ltd., Singapore.

Lim V. (2008) *Creating Paradise T3: Singapore Changi Airport*, SNP International Publishers, Singapore.

Low I. (2009) Looking out for No. 1: Changi Airport may score top marks in the world rankings but there's always room for improvement, *The Sunday Times,* 5 July, p.11.

Low S. P. (1992) *The Just-in-Time Concept to Improving Manufacturing Productivity: Is It Applicable to the Construction Industry?* Construction Papers Series, Paper No. 1, The Chartered Institute of Building, 8pp.

Low S. P. and Chan Y. M. (1997) *Managing Productivity in Construction*, Ashgate Publishing Ltd., Aldershot, UK.

Low S. P. and Chan Y. M. (1996) The application of just-in-time principles to process layout for precast concrete production, *Singapore Management Review*, 18(2):23-40.

Low S. P. and Choong J. C. (2001) Just-in-time management of precast concrete components, *Journal of Construction Engineering and Management*, 127(6):494-501.

Low S. P. and Mok S. H. (1999) The application of JIT philosophy to construction: a case study in site layout, *Construction Management and Economics*, 17(5):657-668.

Low S. P. and Show M. Y. (2008) Facilities design incorporating Just-In-Time Principles for Ramp-up Light Factories, *Facilities,* 26(7):321-342.

Low S. P. and Wu M. (2005) Just-in-time management in the ready mixed concrete industries of Chongqing, China and Singapore, *Construction Management and Economics*, 23(5): 815-829.

MCOT English News (2008) *Dubai outshines Changi airport*, Retrieved on 17$^{th}$ January 2009 from: http://enews.mcot.net/view.php?id=3996.

Ministry of Transport (2008) *Air Transport*, Retrieved on 18$^{th}$ January 2009 from: http://app.mot.gov.sg/data/s_08_03_06a.htm.

Mortimer J. (1986) *Just In Time An Executive Briefing,* IFS Publication Ltd, Bedford, UK.

Newton H. J. (2001) *Statistics*, Texas AandM University, Retrieved on 18th January 2009 from: http://www.stat.tamu.edu/~jhardin/applets/.

Passenger Terminal Today (2008) *Passenger Terminal World – The International Review of Airport Terminal Design, Technology, Security, Operations and Management*, Retrieved on 10th January 2009 from: www.passengerterminaltoday.com.

Schmenner R. W. (1990) *Production and Operations Management*, Macmillan Inc., London.

Schroeder R. W. (1989) *Operations management decision making in the operations function*, McGraw-Hill, Inc., US.

Shingo S. (1982) Study of Toyota Production System from industrial engineering view point, Japan Management Association, Tokyo.

Singapore On The Web (2007) Changi Airport is Best Airport in the World for past 20 years, Retrieved on 18th January 2009 from: http://www.singaporeontheweb.net/previous2007/131007_changiairportis bestairportintheworldforpast20years.htm.

Singapore Sights (2008) Singapore Changi Airport's Terminal 3 – A Green Oasis, Retrieved on 9th January 2009 from: http://www. singaporesights.com/sight/singapore-changi-airport%E2%80%99s-terminal-3-%E2%80%93-green-oasis.

Singapore Tourism Board (2008) Terminal 3's armed for the future; meanwhile Terminal 4 is in the works, Retrieved on 18th January 2009 from: http://www.stbpassport.com/newsbites_sept08_s3.aspx.

Skinner W. (1974) The Focused Factory, Harvard Business Review, 52(3):113-121.

Skytrax (2009) Singapore Changi Airport Review, Retrieved on 9th January 2009 from: http://www.airlinequality.com/Airports/Airport_forum/sin. htm.

Straits Times (2008) Changi staff commended for customer service, Retrieved on 20th January 2009 from: http://www.straitstimes.com/ print/ST+Forum/Story/STIStory_184249.html.

Swanson C. A. and Lankford W. M. (1998) Just-in-Time Manufacturing, Business Process Management Journal, 4(4):333-341.

Thomas B. (1987) Applicability of Just-In-Time Techniques in the Administrative Area, Unpublished PhD Thesis, University of Nebraska.

Travel Daily News (2008) *Changi airport is shopping haven for business traveler,* Retrieved on 11th February 2009 from: http://www.traveldailynews.com/pages/show_page/27178.

Walker R. (2001) *Meet the new airport: temple, mall, design hub*, Christian Science Monitor, Retrieved on 10th January 2009 from: http://www.csmonitor.com/2001/0823/p18s1-altr.html.

# INDEX

## A



## B



## C

## D



## E



## F

## G



## H



## I



## J



## L



## M